Lemuel Bolton Bangs

Atlas of Syphilis and the Venereal Diseases

including a brief treatise on the pathology and treatment

Lemuel Bolton Bangs

Atlas of Syphilis and the Venereal Diseases
including a brief treatise on the pathology and treatment

ISBN/EAN: 9783743359604

Printed in Europe, USA, Canada, Australia, Japan

Cover: Foto ©berggeist007 / pixelio.de

Manufactured and distributed by brebook publishing software (www.brebook.com)

Lemuel Bolton Bangs

Atlas of Syphilis and the Venereal Diseases

ATLAS
OF
SYPHILIS
AND THE
VENEREAL DISEASES

INCLUDING

A BRIEF TREATISE ON THE

PATHOLOGY AND TREATMENT

BY

PROF. DR. FRANZ MRAČEK

of Vienna

AUTHORIZED TRANSLATION FROM THE GERMAN

EDITED BY

L. BOLTON BANGS, M.D.

Consulting Surgeon to St. Luke's Hospital and the City Hospital, New York; late Professor of Genito-Urinary Surgery and Venereal Diseases, New York Post-Graduate Medical School and Hospital

With 71 Colored Plates

PHILADELPHIA

W. B. SAUNDERS

925 WALNUT STREET

1898

LIST OF ILLUSTRATIONS.

EDITOR'S PREFACE.

THOROUGHNESS has always been granted as one of the chief attributes of German teachers. The present volume, in the opinion of the editor, shows another trait no less characteristic, but which is perhaps not so generally conceded. It is a certain strong practical sense in selecting the material and presenting it in the form best fitted to the needs of those for whom it is intended. On the latter point the author has defined the purpose of his work in his preface. Making some allowance for local customs and conditions, especially in the matter of treatment, the book will, it is believed, prove a thoroughly practical manual for the every-day use of the practising physician.

The translator has endeavored to follow the German text as faithfully as possible, and has ventured to depart from the original only when the needs of the language appeared to justify such a departure. The translation will therefore be found to be a conscientious reproduction of the author's ideas in concise, smooth, and graphic phraseology.

For the sake of convenience the names of drugs and certain technical terms have been made to conform to the current English expressions, and in the prescriptions the equivalent amounts in apothecary's weight have been added in parentheses after the number of grams.

A short index has been added, which is to be regarded rather as a supplement to the table of contents and the list of illustrations.

PREFACE.

In complying with the publisher's request, made to me two years ago, to prepare an Atlas of Venereal and Syphilitic Diseases, I have endeavored to produce a book which should be within the reach of a wider circle of readers, as both the scope and the cost of the pictorial works hitherto published on the subject necessarily restrict their influence to a comparatively small number. To carry out this plan it was found necessary to select those diseases which are of the most frequent occurrence and greatest practical importance, omitting such as interest only the specialist. The same principle controlled the composition of the text. The clinical material is drawn chiefly from my ward in the K. K. Rudolfsspital in Vienna. I have to express my thanks to Dr. Braun, Director of the Northern Austrian Foundling Asylum, for two cases of hereditary syphilis; to (Prof.) Dr. Bergmeister, "Primararzt," for two cases of syphilitic eye-disease, and to (Prof.) Dr. E. Lang for a case of disease of the hairy scalp.

Mr. Schmitson mastered the difficult problem of interpreting and reproducing the various clinical pictures in an amazingly short time, and has turned out a truly admirable set of water-colors, the reproduction of which by the publishing house has been done in the most exemplary manner.

In the work of sifting the case-histories and preparin the material I have been most ably and zealously aide by my assistant, Dr. Grosz.

To all these gentlemen I wish to express my heart thanks, and I trust that the present work may meet wit a kind reception among those for whom it is intended

DR. MRACEK.

VIENNA, November, 1897.

CONTENTS.

PLATE 1.

Sclerosis in the Coronary Sulcus of the Penis.

Present Condition.—On the dorsal aspect of the neck a sclerosis measuring about 1 cm. in diameter; the surface is necrotic, the base and surrounding tissue moderately infiltrated. Perceptible swelling in the inguinal glands; on the trunk a pale, slightly raised eruption.

Sch. J.; admitted Nov. 24, 1895. Says he had his last coitus Oct. 4th; the sore on the penis developed four weeks later. Has always been healthy.

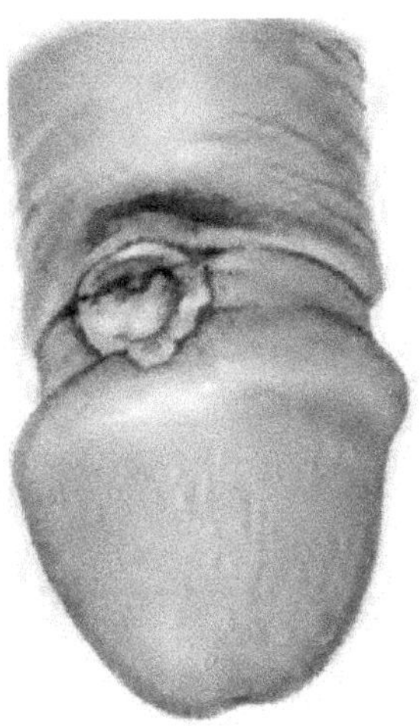

Lith.Anst.v. F. Reichhold. München

PLATE 2.

Sclerosis in the Pubic Region.

Present Condition.—In the pubic region there is a sclerosis about the size of a hazelnut which is quite deep; the floor is covered in part with pus and in part with pale granulations; the edges are sharp, tough and infiltrated, and surrounded by a zone of inflammation. The inguinal glands are greatly enlarged; the axillary and cervical, only moderately so.

M. T., 34 years old, works in a gas-factory; admitted June 7, 1896. In the beginning of May a glowing piece of coke fell on the patient's bare breast; in attempting to shake it out of his clothing he burnt himself in the pubic region. Two weeks later he had coitus. He had no suspicion of the true nature of his disease. On June 20th a dark-red, papular syphilide appeared on the body.

Local treatment and twenty-five inunctions effected a cure.

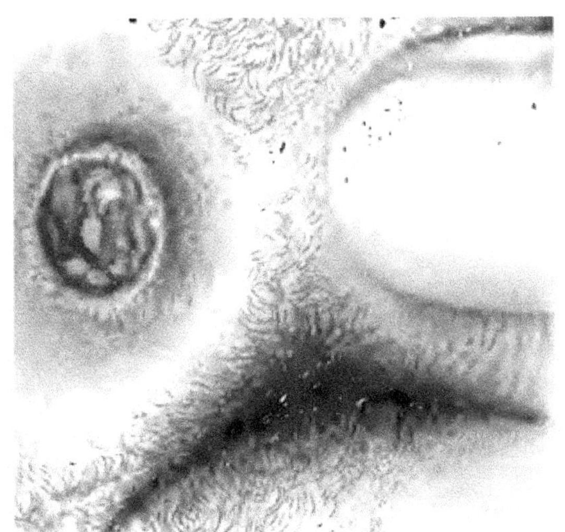

PLATE 3.

Sclerosis on the Anterior Surface of the Scrotum.

Present Condition.—On the scrotum, below the angle formed by the penis (penoscrotal angle), is seen a sclerosis a little larger than an almond. The surface is ulcerated, the base and edges infiltrated. The rest of the genital region, as well as the skin on the trunk and extremities, is covered with recent papules as large as lentils. The older of these papules already show desquamation on the surface.

W. A., 28 years old, mail-driver; admitted Nov. 1, 1895. The patient says he paid no attention to the sore on the scrotum at first. It began to be more noticeable four weeks ago; the eruption appeared only six days ago.

After local treatment with gray plaster and a course of twenty-five inunctions the sclerosis healed completely, the infiltration at the base softened, and the eruption disappeared. The patient was discharged on Dec. 2d, after thirty-two days' treatment.

PLATE 4.

Sclerosis on the Right Labium Majus.

The right labium majus is moderately swollen and edematous. On the external surface of the lowest segment there is a hard lump about as large as a penny; the crater-like center is covered with pus and discharges quite freely.

In addition to the sclerosis described, the patient exhibits a macular syphilide on the trunk and a lenticulo-papular eruption on the thighs and nates. The inguinal, as well as the cervical and axillary lymph-glands on both sides of the body are involved.

K. C., 20 years old; admitted Aug. 16, 1896. Her first venereal attack, which she says began three weeks ago (?).

After fifteen inunctions the sclerosis healed, the edema disappeared, the eruption became less angry (paler), and the infiltration about the site of the sclerosis as well as the glandular swelling diminished.

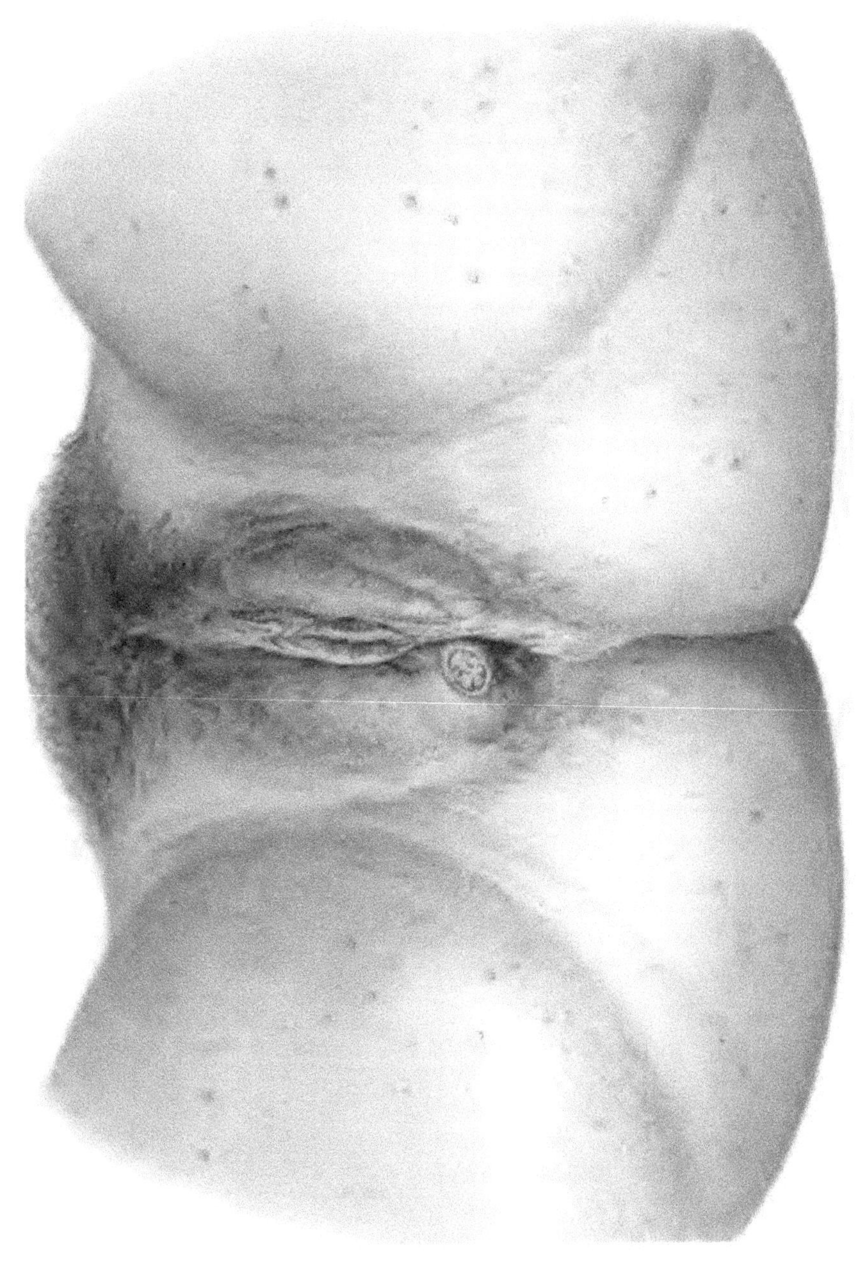

PLATE 5.

Ambustiform Sclerosis and Indurative Edema of the Left Labium Majus.

Present Condition.—The entire left labium majus shows a livid discoloration and is considerably swollen and indurated. About the middle is an ulcer with hemorrhagic floor and slightly eroded edges, resembling a wound made by a red-hot instrument (sclerosis ambustiformis). The inguinal glands, especially on the left side, are swollen, and the axillary and cervical glands slightly enlarged. Patient had been suffering from insomnia for a week. Later in the course of the disease a roseola appeared on the trunk.

H. M., 21 years old, cashier; admitted Oct. 13, 1896. The patient says she noticed her condition only a week ago. Last coitus seven weeks ago.

After the use of inunctions the eruption disappeared, the glandular swelling diminished, the sclerosis healed, and the indurative edema was reduced to an elastic thickening of the labium.

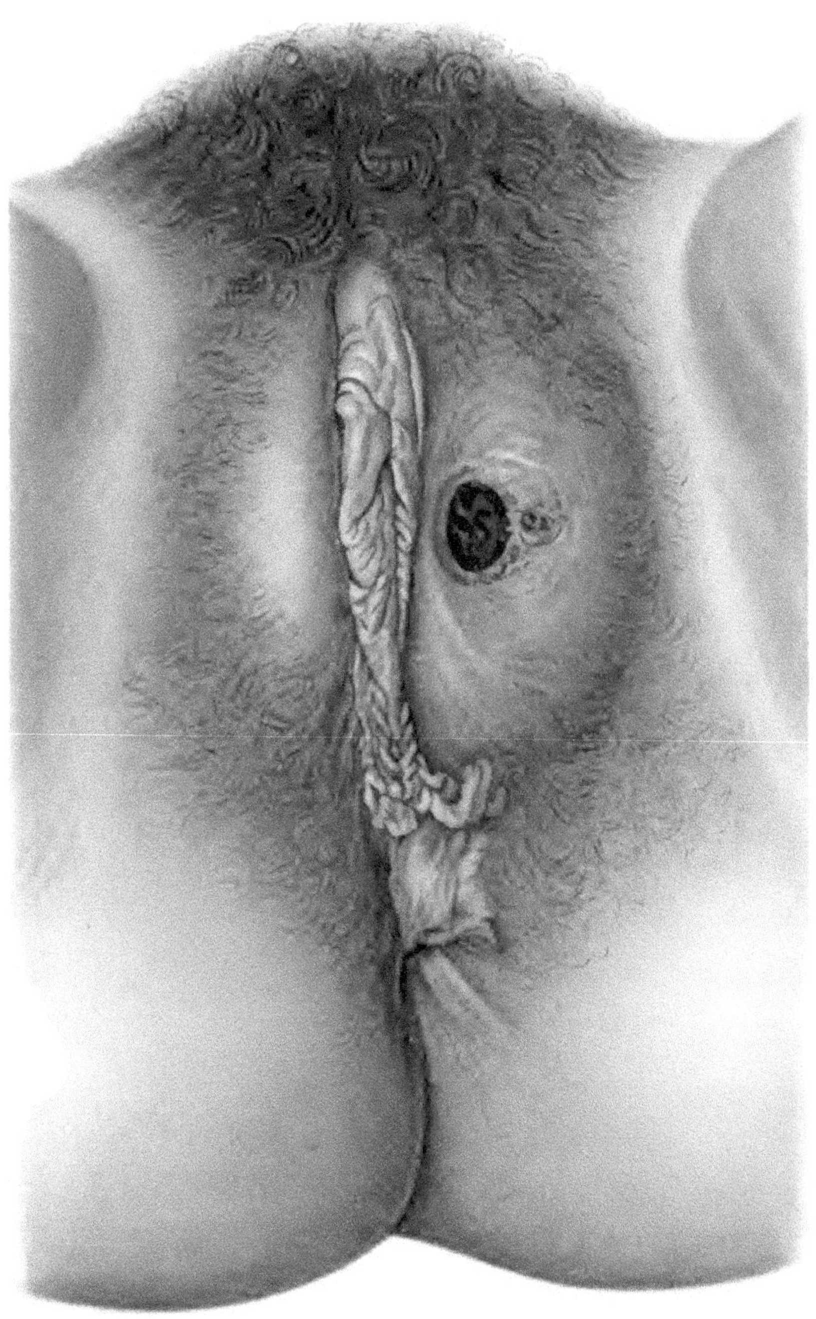

Lith Anst v. F Reichhold, München

PLATE 6a.

Ulcerative Sclerosis in the Vaginal Portion of the Cervix.

The vaginal segment enlarged as a whole; the os slightly contracted by scar-tissue. On the anterior lip, close to the os, is a slightly raised sclerosis, the floor of which presents a diphtheritic appearance and is marked in places by small hemorrhages. On palpation a lump as hard as cartilage can be plainly made out in the vaginal portion. There is a shallow erosion on the posterior lip.

B. A., 22 years old. The patient has given birth to one child. She first became aware of her disease when a sclerosis, similar to the one described, appeared on one of the labia. She knows nothing of the sclerosis in the vaginal portion. Last coitus seven weeks ago, last but one eighteen months ago.

Later in the course of the disease a macular syphilide made its appearance. Inguinal, cervical, and axillary glands enlarged.

Inunction treatment.

The sclerosis was excised and examined under the microscope.[1]

[1] See *Vierteljahrschrift für Dermatologie*, 1881, page 57 *et seq.*

PLATE 6b.

Sclerosis in the Vaginal Portion of the Cervix.

Present Condition.—Vagina pale and distended; secretion scanty. Cervix large and cylindrical. On the anterior lip is a circular, sharply circumscribed ulcer about as large as a penny, with purulent floor. Surrounding tissue much inflamed. Glands palpable everywhere.

P. C., 43 years old, prostitute; admitted April 15, 1896. Says she has been ill six days.

Treatment.—White-precipitate ointment. Cicatrization May 4, 1896.

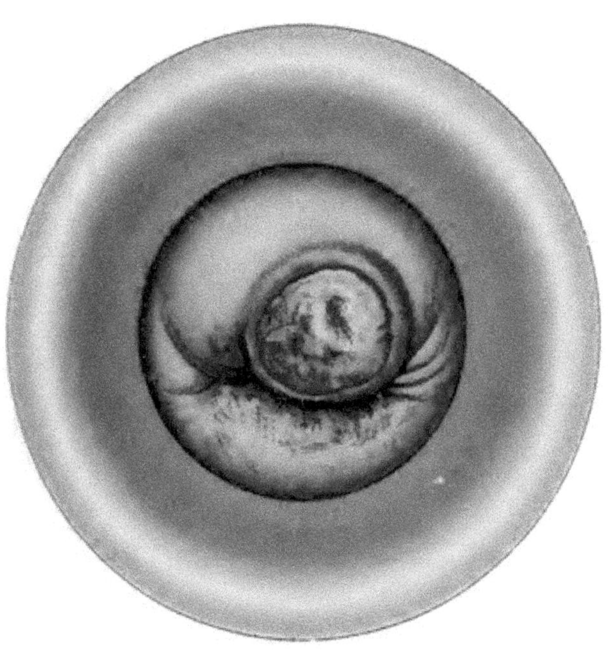

a

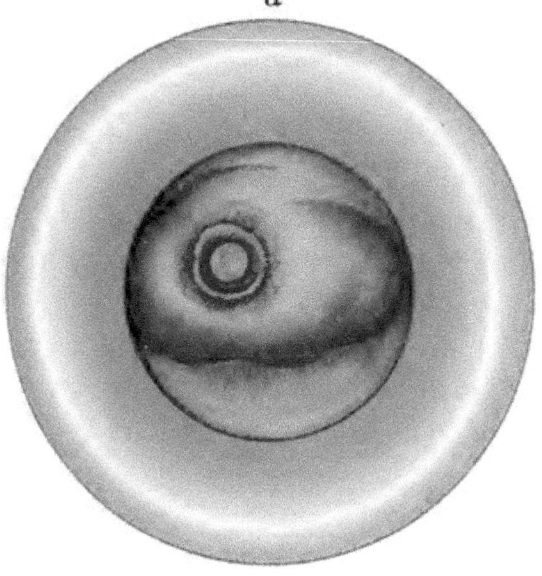

PLATE 7.

Two Ulcerative Scleroses on the Os Uteri.

The cervix as a whole is swollen and edematous, and shows several cicatricial contractions, the result of former confinements. On symmetrical areas of the anterior and posterior lips are two sharply circumscribed scleroses, with raised edges and purulent floor. The surrounding tissue is very hard to the touch. The secretion is serous in character and not particularly copious.

G. M., 24 years old. The patient was not aware of her disease, but came to be treated for papules in the vestibule. In addition to the symptoms described, the patient had a slight macular eruption and glandular enlargement.

Patient was subjected to inunctions.

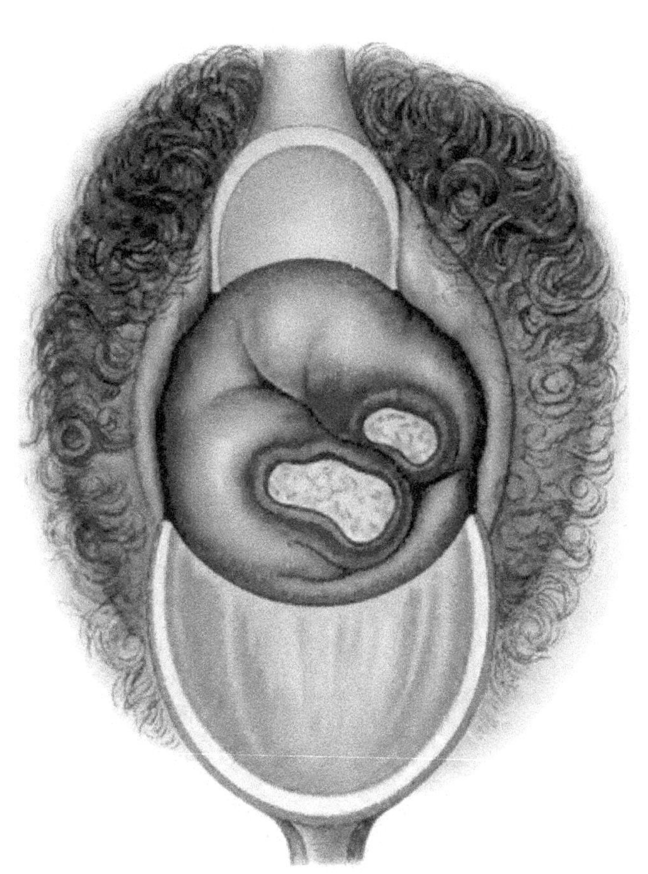

PLATE 8.

Sclerosis on the Lower Lip.

Present Condition.—On the lower lip, to the right of the middle line, is an ulcer, about 1.5 cm. in thickness, the center somewhat depressed, with slightly raised edges. There is some infiltration and the ulcer is breaking down.

The patient, K. M., 21 years old, does not know the cause of this ulcer, which appeared about four months ago and has been steadily growing larger. The only other symptom is swelling of the submaxillary glands; the cervical, axillary, and inguinal glands are also enlarged; the genitalia are intact.

The ulcer healed after the application of gray plaster and about thirty inunctions.

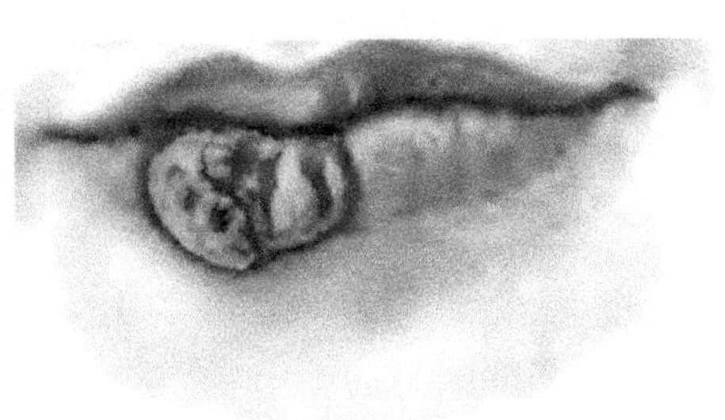

Lith.Anst.v. F Reichhold, München

PLATE 9.

Sclerosis at the Right Angle of the Mouth.

N. P., 29 years old, blacksmith; admitted Oct. 30, 1895. Eight weeks ago the patient first noticed a sore at the right angle of the mouth, which slowly but steadily increased in size. Five weeks ago the right cheek and the right submaxillary region began to swell. The patient does not know the cause of his disease; but he says that at the time it began there was a man in the shop where he works who was suffering from a chancre. For the past two weeks he has had severe nocturnal headache.

Present Condition.—In the mucous membrane of the right angle of the mouth an oval, cup-shaped ulcer, over ½ cm. long, covered with pus. The right side of the face is swollen. The submental and submaxillary lymph-glands are enormously enlarged. The other lymph-glands are also involved, but not to the same degree. Maculopapular syphilide on the trunk and extremities.

Cured after twenty-five inunctions.

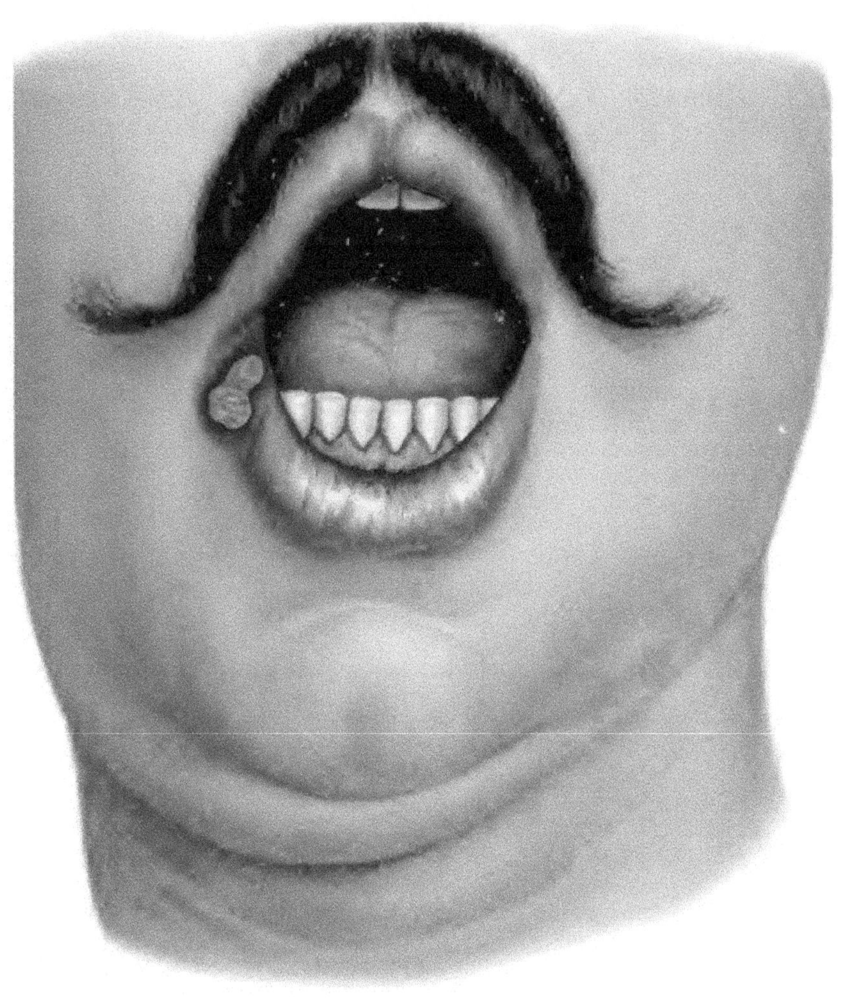

Lith.Anst.v. F. Reichhold, München

PLATE 10.

Sclerosis of the Tongue.

Fr. Th., 25 years old, working-woman; admitted Oct. 22, 1896. An ulcer of unknown origin appeared on the tongue a month ago.

Present Condition.—The right margin of the tongue exhibits a hard nodule about as large as a bean, partly embedded in the substance of the tongue and partly rising above its level. In the center is a flat, oval ulcer, the floor of which is covered with a grayish-white secretion. The submaxillary glands on the right side are enlarged to the size of a pigeon's egg and painful on pressure. Cervical and axillary glands can be plainly felt. Genitalia normal. A few scattered macules on the trunk.

Nov. 2. Papules have made their appearance between the macules. Headache.

Subsequent Course.—After twenty inunctions complete regeneration of the nodule on the tongue, and disappearance of all other specific symptoms.

Nov. 28. Discharged cured.

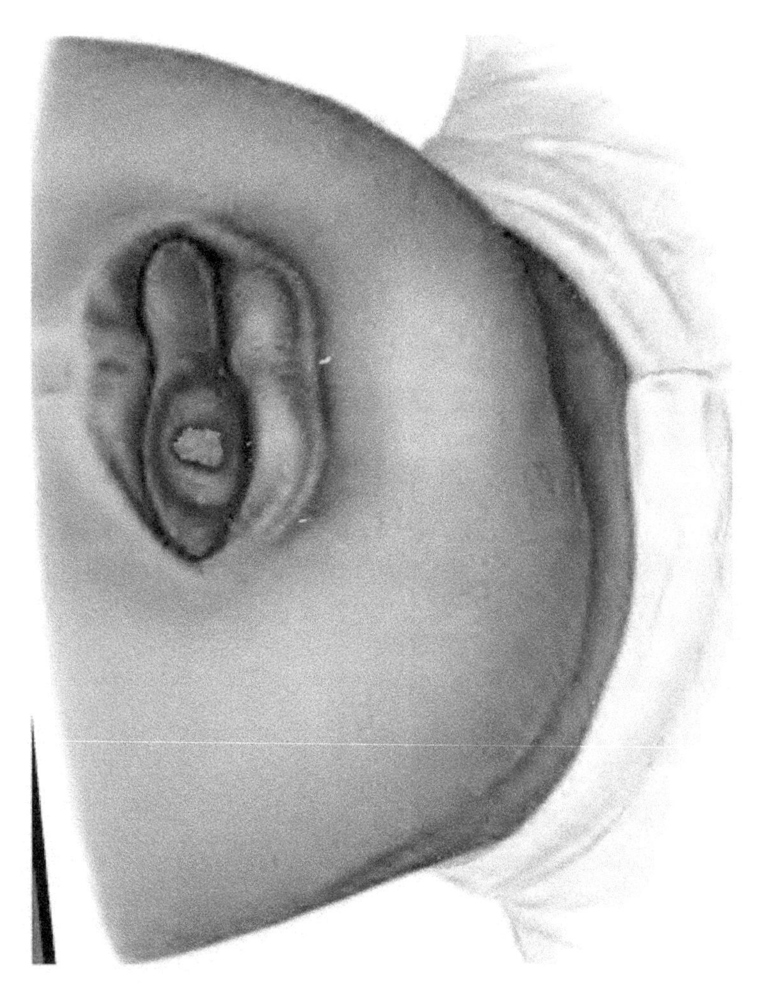

PLATE 11.

Sclerosis of the Right Tonsil.

W. W., 26 years old, roofer; admitted June 29, 1896. Patient says he has always been healthy; since May 13th has noticed a swelling of the right tonsil and pain on swallowing.

Present Condition.—The right tonsil is larger than a pigeon's egg and reaches almost to the middle line, forcing the palatoglossal and palatopharyngeal arches far apart. It shows considerable infiltration and is covered with partly necrosed ulcers. The mucous membrane of the adjoining tissues is inflamed and slightly swollen. The inflammation extends as far as the uvula on the left side, and in front as far as the anterior border of the soft palate. Under the angle of the jaw, on the right side, is a slightly movable tumor, about as large as a hen's egg, which corresponds to the submaxillary gland. The middle cervical glands and the supraclavicular glands on the right side are as large as beans or hazelnuts, easily movable, but not painful. The left middle cervical and the axillary and inguinal glands are also palpable, but not swollen.

July 1. A roseolar eruption over the entire trunk. The necrotic covering of the sclerosis has fallen off.

July 5. The syphilide and the glandular swelling increase. Granulation is forming in the sclerosis.

Treatment.—Gargles. The sclerosis was painted with tincture of iodin. Twenty inunctions.

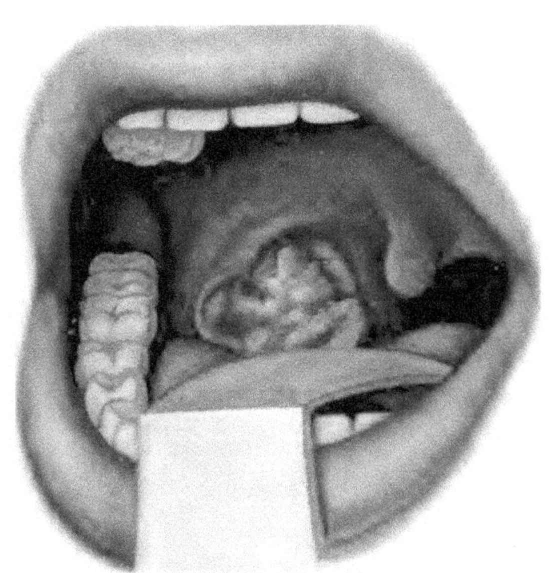

PLATE 12.

Indurative Syphilitic Edema.

J. S., 20 years old, butcher. Under treatment from Oct. 24 to Dec. 9, 1890.

Two months ago the patient contracted a sore at the lower margin of the prepuce. Within a week the entire penis was swollen and inflamed. Three weeks ago the scrotum began to swell. The prepuce, which accordingly has been swollen for seven weeks, could not be pushed back by the patient, so that he can give no account of the subsequent course of the wound.

Present Condition.—Edematous phimosis of the prepuce. Lymphangitis and edema of the skin of the entire penis. Indurative edema of the scrotum with superficial erosions. Bilateral inguinal lymphadenitis. General disease of the lymphatic system. Both tonsils are enlarged and covered with diphtheritic papules. Psoriasis plantaris. On the trunk a papular syphilide in process of regeneration. Length of the penis, 13 cm.; circumference taken about the center, 11.5 cm.; circumference of the scrotum, measured in the sagittal direction from the root of the penis to the perineum, 26 cm.; circumference of the scrotum, measured in the frontal direction from one inguinal fold to the other, 30 cm. The skin of the scrotum is dark red, hot and infiltrated. On the scrotum and on the penis the skin is coming off in single, attenuated layers; a few erosions on the scrotum. The integument of the scrotum is so thick and infiltrated that the testicles cannot well be made out by palpation.

Treatment.—Irrigation of the preputial sac. Compresses wet in aluminum acetate solution. On Oct. 30th inunctions were begun, and the swelling and infiltration of the skin of the scrotum and penis began to subside. Reposition of the prepuce revealed a shallow scar, as large as a pea, on the inner surface of the prepuce and on the lower aspect of the glans. Circumcision. Entire disappearance of all the symptoms after thirty inunctions.

PLATE 13.

Recent *Macular* Eruption (Roseola) over the Entire Surface of the Body.

The skin of the entire body is thickly covered with red spots, of a darker color on the dependent portions of the body than about the trunk, neck, and face; the spots are not shiny and do not desquamate. The soles of the feet and palms of the hands exhibit a brownish, papular eruption (psoriasis plantaris et palmaris).

The prepuce is indurated and shows the scar of a sclerosis. The inguinal, cervical, and axillary glands are enlarged.

B. L., 23 years old, laborer; admitted Aug. 4, 1897. The patient states that he was discharged from a hospital a month ago, after the sore on his penis was cured, without being subjected to any general treatment. He is unable to give the exact date of his infection (a little more than two months ago).

After being treated with inunctions for thirty days the patient was discharged cured.

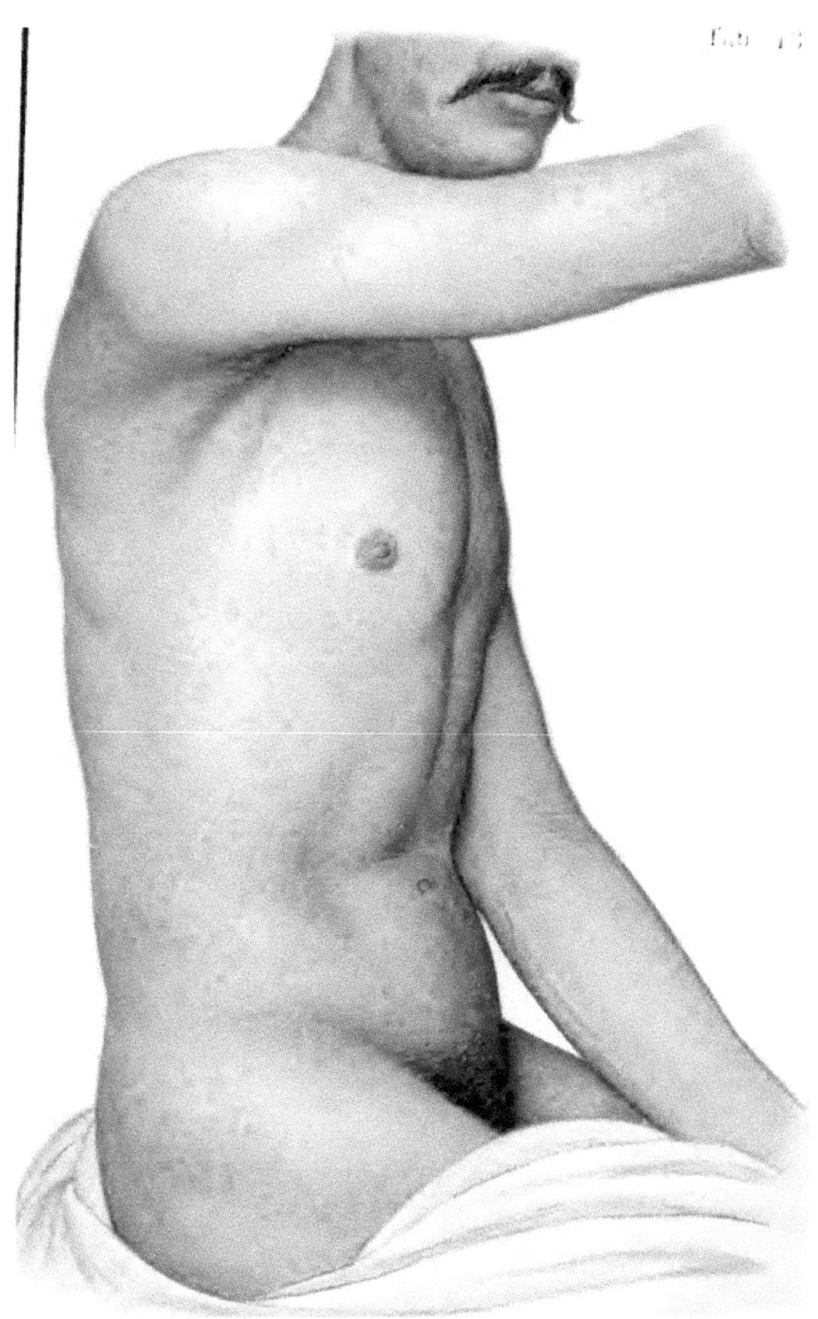

PLATES 14, 14a.

Large Macules Mixed with Papules, Scattered over the Entire Body.

H. S., 19 years old; admitted Jan. 27, 1896. The history, as given by the patient, is extremely inaccurate, and amounts to this, that he has been ill three months and has done nothing so far to cure himself.

Present Condition.—A livid, somewhat infiltrated scar, the result of the primary lesion, on the outer layer of the prepuce. Moist papules in the coronary sulcus of the penis and about the anus. An abundant macular syphilide scattered over the trunk and extremities. Here and there among the macules are seen large, shiny papules, their reddish-brown color contrasting with the livid hue of the macules. The soles of the feet exhibit dirty yellow papules (psoriasis plantaris). Alopecia and slight desquamation of the hairy scalp. On the forehead a maculo-papular syphilide. Both tonsils are enlarged and covered with coalescent, suppurating ulcers (papules).

Treatment.—Labarraque's solution externally. Inunctions.

Black Plate (Pl. 14a): Dorsal view, showing the widespread large macular syphilide. *Colored Plate:* Right forearm, with the same large macular syphilide.

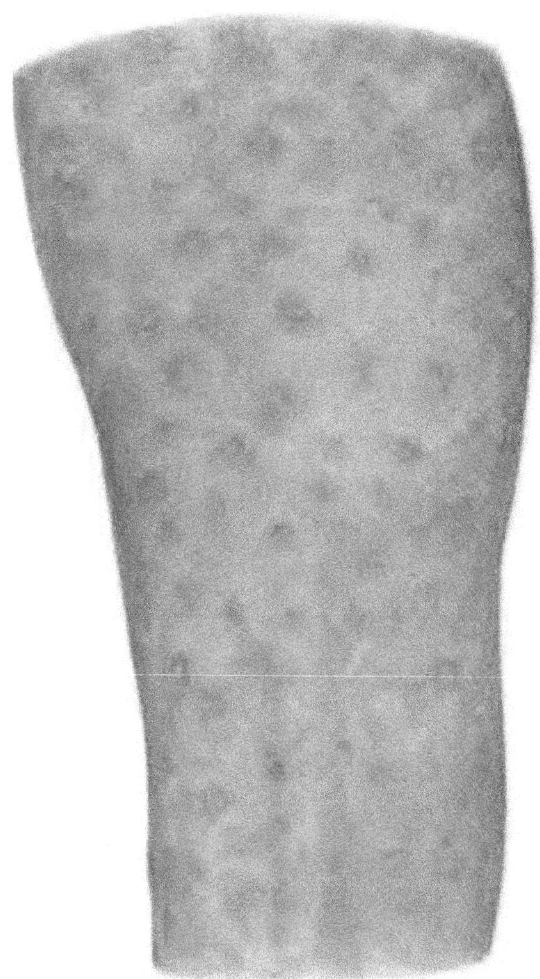

Tab.

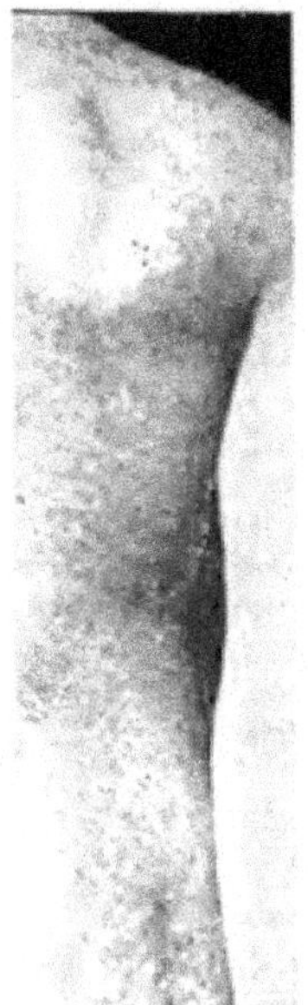

PLATE 15.

Erythema Figuratum.

H. J., 42 years old, saloon-keeper; admitted July 12, 1896. Became infected eighteen months ago, at which time he was subjected to inunctions at this hospital. Since then he has had several relapses, taking the form of papules on the mucous membranes, which were treated with milder remedies (internal administration of mercurial preparations, potassium iodid, and external applications of chromic acid). He says the eruption appeared a week ago. The patient is a moderate drinker and smoker.

Present Condition.—On the dorsum of the penis is an infiltrated scar as large as a penny. The lymph-glands which can be felt are spindle-shaped. About the middle of the right border of the tongue is a papular efflorescence about as large as a pea, with ulcerated surface. Otherwise the mucous membranes are free from disease. The skin of the trunk and of the upper extremities is covered with a pale-red exanthema, arranged in sinuous figures formed by the confluence of circular eruptions. The rash is distributed over the extensor surfaces of the arms and over the trunk, being more distinct on the sides of the thorax than on the back and buttocks. The graceful figures stand out very plainly after the naked body has been exposed to the air. The patient complains of headache, worse at night. Psychical condition intact. Pupillary reaction and tendon-reflexes normal.

Treatment.—Potassium iodid } āā 1.00 (gr. xv).
Potassium bromid }

To be taken at night.

Cured.

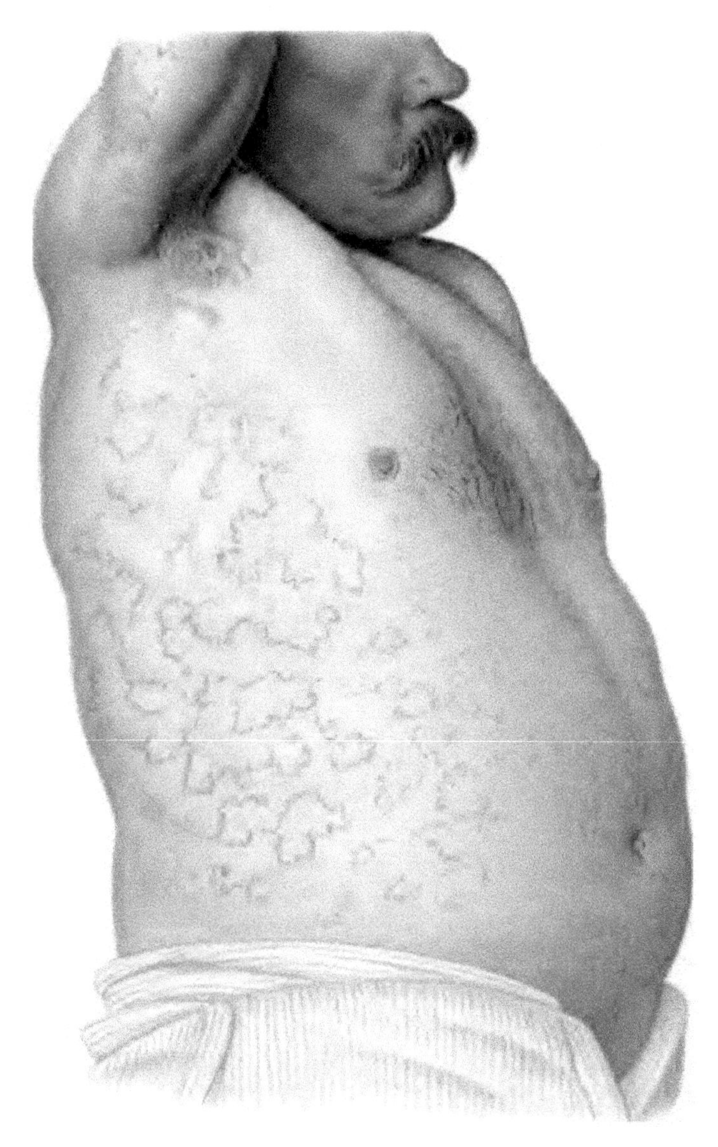

PLATE 16.

Erythema Figuratum.

(Erythème circinée—Fournier.)

E. B., 26 years old, clerk; admitted Dec. 21, 1896. The specific infection occurred in February of the present year; the patient was treated with injections at the time. For the past two weeks he has had difficulty in swallowing; he has no knowledge of any rash.

Present Condition.—Eroded papules on both tonsils. The latter, as well as the pillars of the fauces and the posterior wall of the pharynx, are swollen and inflamed. The entire lymphatic system is diseased. A pale-red eruption appears distributed almost symmetrically over the skin of the trunk and extremities. The individual patches of the eruption are circular in shape and vary in size from a penny to a dollar. By the confluence of adjoining patches the eruption assumes the form of festoons or garlands, the general arrangement of the figures corresponding to the slant of the ribs on the back, chest, and sides of the thorax. The face, palms of the hands, and soles of the feet are free.

Cured after twenty-five inunctions.

Tab. 16.

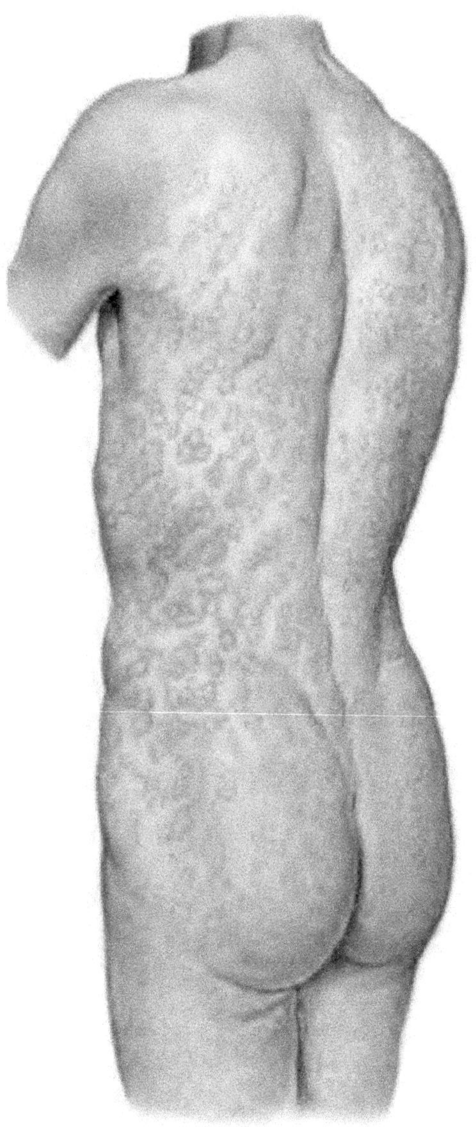

Lith. Anst. v. F. Reichhold, München

PLATE 17.

Syphilitic Papules Distributed over the Entire Body.

L. M., 30 years old, laborer; admitted July 4, 1897. The patient says he performed his last coitus two months ago. Immediately afterward he noticed a sore on the foreskin. The eruption he noticed eight to ten days ago. Has had no treatment so far.

Present Condition.—The skin of the entire body is of a brownish hue and covered with copper-colored nodules as large as lentils. The eruption is situated mostly at the sides of the thorax, on the abdomen, and on the flexor surfaces of the extremities. Most of these papules already show a whitish discoloration of the epidermis at the apex, which can be removed in some of them by very light abrasion with the finger-nail. All the lymphatic glands are moderately enlarged. On the dorsal aspect of the neck of the penis is a livid, recently healed, sclerosis, surrounded by a good deal of induration and infiltration. The mucous membranes, palms of the hands, and soles of the feet are not involved. A few papules are seen on the face, at the roots of the hair. The hairy scalp shows slight seborrhea, but no distinct papules.

Treatment.—Antiseptic mouth-wash; baths; twenty-five inunctions. Cured.

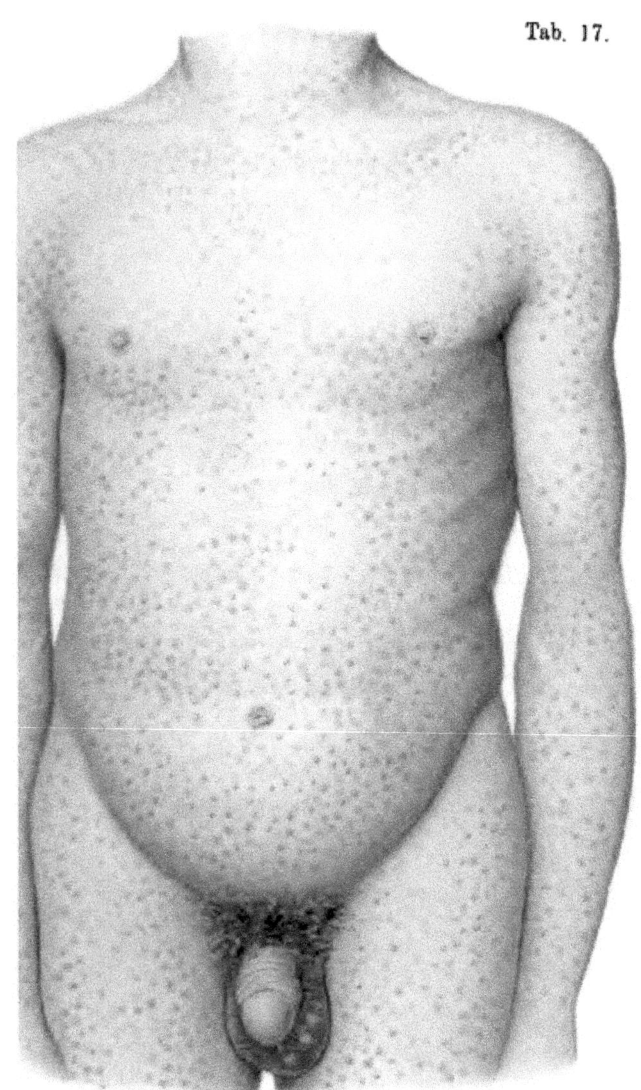

Lith. Anst. v. F. Reichhold, München

PLATE 18.

Papulopustular Syphilide. Jaundice.

M. S., 24 years old, nurse. Under treatment from Feb. 11 to March 20, 1897. The patient says she had jaundice five weeks before entering the hospital; the eruption appeared only during the last week. She has suffered a good deal lately with frontal headache, especially at night; also complains of sore throat. Last coitus three months ago.

Present Condition.—On the right labium majus a sclerosis as large as a hazelnut, with ulcerated surface. General glandular enlargement. The skin and mucous membranes, wherever visible, present an intense yellow discoloration and are thickly covered with innumerable papules, varying in size from a millet-seed to a lentil. Here and there, especially on the back and in the intermammillary region, are numerous pustules covered with hemorrhagic crusts. Recent eruption of psoriasis plantaris. Mucous membrane of the mouth intact. The face wears an expression of suffering. Violent headache.

Treatment. — Labarraque's solution locally. Antiseptic mouth-wash.

Feb. 15. Patient very much prostrated; complains of violent headache, especially at night; temperature normal; percussion and palpation bring out no enlargement of liver or spleen. Inunctions.

After fifteen inunctions the jaundice disappeared entirely, the patient began to feel better, and the specific symptoms began to subside, the papules and pustules being replaced by pigmentation. After thirty inunctions the patient was discharged cured on March 20th.

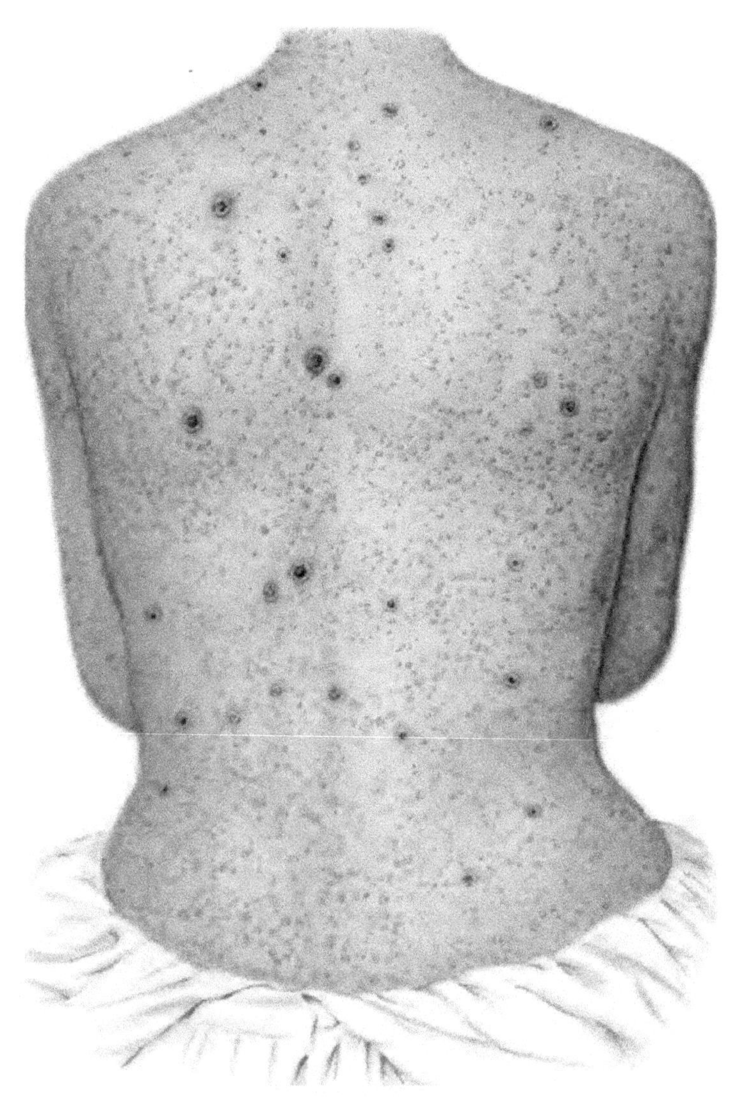

PLATE 19.

Small, Aggregated Papules (Lichenoid Syphilide, Relapsing Form).

S. A., 19 years old, seamstress; admitted Jan. 24, 1896. Patient was treated for syphilis in Sept. and Oct., 1895. The present trouble developed a month ago.

Present Condition.—On both labia majora and in the genitocrural fold on both sides are seen papules as large as peas and elevated above the surrounding level. Over the sacrum and on the nates, small aggregated papules. Some of the papules are desquamating, others show a reddish-brown pigmentation. Similar patches are seen over the knee-joint, on the lower abdomen, and at the back of the neck. Moderate itching in the affected parts.

Cured after twenty inunctions of 5 g. (ʒjss) each.

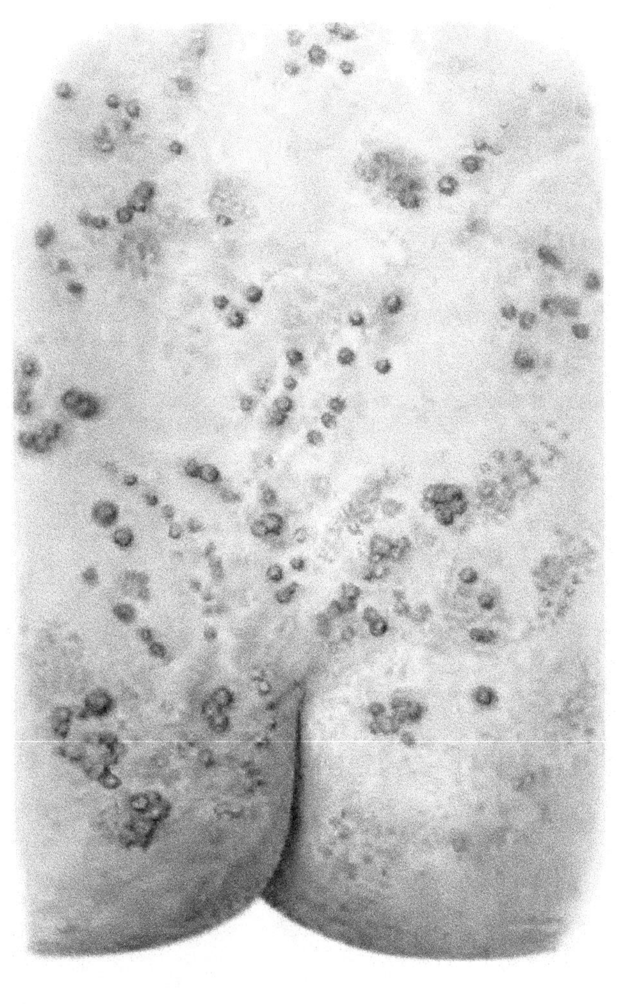

PLATE 20.

Papulosquamous Syphilide.

T. J., 34 years old, hostler; admitted March 27, 1896. Toward the end of June, 1895, four weeks after coitus, the patient had an ulcer on the penis. He then had a private physician, and was treated with yellow-precipitate ointment and sublimate baths locally. After the ulcer healed the patient again indulged in sexual intercourse; three weeks later an eruption appeared over the entire body (from his description a macular and papular syphilide). On Aug. 20, 1895, he applied for admission to the general hospital, and was treated until Oct. 31st. He was subjected first to twenty-three inunctions with calomel, and then to forty-seven inunctions with gray ointment. On Nov. 24th patient applied for admission to the Rudolfspital. He then complained of pains in the head, in the epigastrium, and in the thorax. The patient was pale; the skin of the trunk was thickly covered with livid and brownish patches as large as a pea, the remains of an old syphilide; the epidermis over most of the patches is marked by fine furrows. Inguinal, axillary, and cervical glands enlarged. Incipient leucodermia colli. Pharyngitis. The internal organs were normal. On Dec. 11th a tassel-shaped eruption appeared on the forehead. On Dec. 18th the tassel disappeared, leaving a livid spot, and appeared in another place. At the same time several papules appeared, disposed in a circle about the left nostril, and also another papular eruption on the throat. The gums are eroded. Gingivitis. Seborrhœa capitis. Alopecia. General neurasthenia. On Dec. 24th single pustules appeared on the head. On Jan. 2d the patient felt completely cured, and was, at his own request, discharged from the hospital. On March 27, 1896, he again asked to be admitted to the hospital. Examination revealed the following condition : in addition to the remains of the old syphilide, which existed during his former stay at the hospital, the entire body is now covered with a new eruption; the center of the papules, which vary in size from a pea to a bean, consists of a raised, whitish scab, while the margin is of a light-red color. In some places the central scab has fallen off, leaving only a livid, red patch, slightly raised above the level of

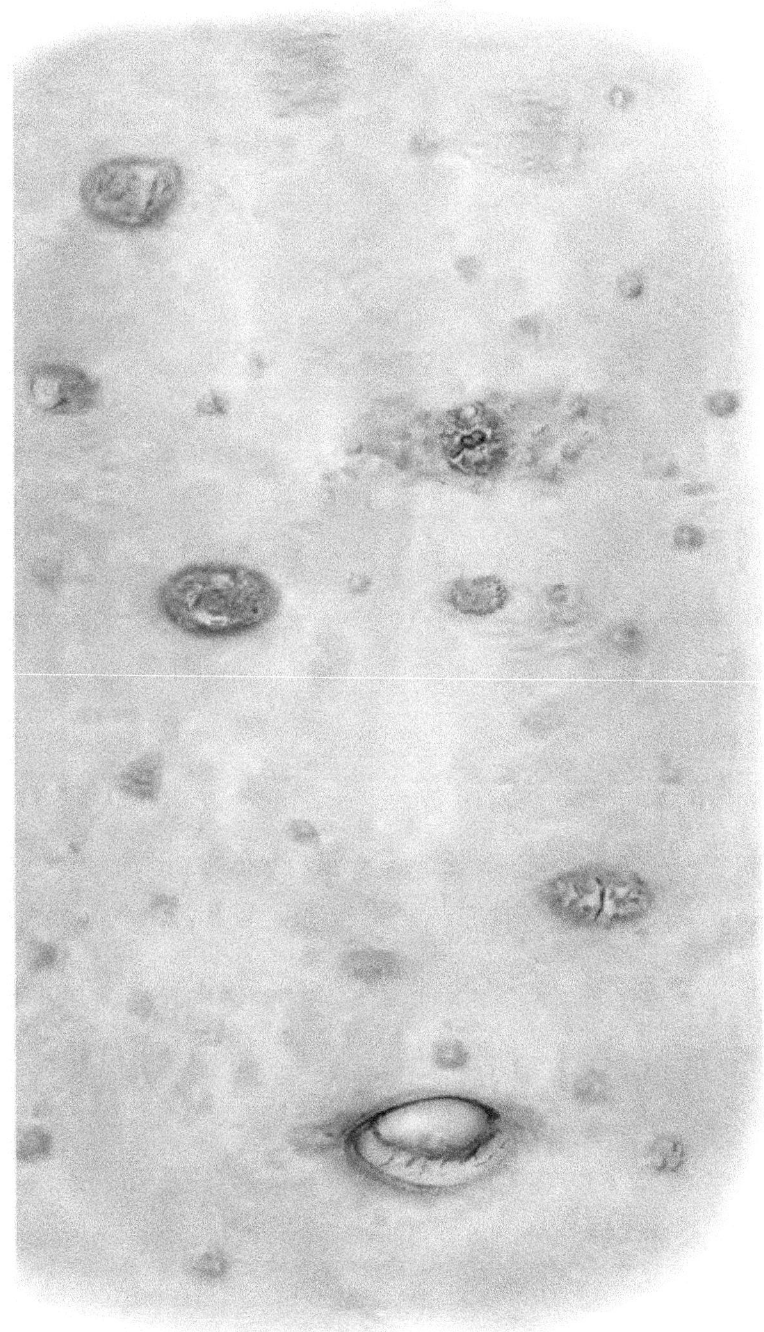

Lith.Anst.v. F. Reichhold, München

the skin. The center of these patches is badly degenerated and converted into a hemorrhagic wound, and around it are grouped thick clusters of small, miliary papules. The isolated, small nodules show a slight whitish discoloration at the apex, but no desquamation. In addition, the remains of the old syphilide are seen here and there (at the top of the illustration, for instance) in the form of light, reddish-yellow patches of epidermis marked by delicate striations. The lower part of the navel is occupied by a phagedenic, moist papule. The general color of the skin is dirty yellow, with here and there irregular lighter areas, the remains of former eruptions. Ulcerating papules on the scrotum; papules on the mucous membrane of the mouth and of the lower lip, and on both tonsils.

Treatment.—Mixed. Cure.

PLATE 21.

Papular, Orbicular Syphilide.

M. E., servant-girl, 26 years old; admitted Oct. 1, 1895. The patient became aware of her syphilitic disease by accident, a police-surgeon calling her attention to it on the occasion of her being incarcerated, although she knew that an eruption had begun to develop in the genital region for a year, and on the lower limbs and on the neck for the last five months. She, however, attached no importance to it and did nothing for it. She has never given birth, menstruates regularly, and says that she had her last coitus more than a year ago.

Present Condition.—In both genitocrural folds, at the edges of both labia majora, and about the anus are seen proliferating papules, some of which have run together. Part of the trunk is covered with the pigmented remains of a papulo-serpiginous syphilide, the rest by an irregular (figurate) syphilide made up of large macules. On the legs is a lichenoid, brownish-yellow syphilide, arranged in groups. On the inner aspect of each thigh is an elliptical, orbicular syphilide. The margin is composed of small, lichenoid papules surrounded by a coppery halo; the center is brown, with a shade of gray, and the epidermis is in part undergoing desquamation. Above, the line of small papules is irregular, so that the margin of the entire ellipsoid figure appears broken. In addition, all the lymph-glands are swollen; the mucous membranes are free.

After local treatment with sublimate and gray plaster, and forty inunctions, the syphilide disappeared, leaving some pigmentation. The patient was discharged cured after having been under treatment sixty-five days.

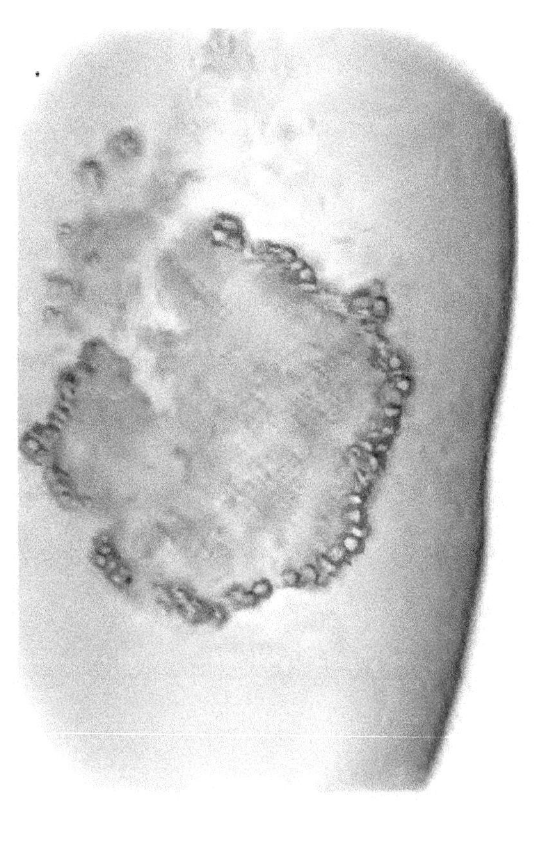

PLATES 22, 23.

Irregularly Distributed Papular Syphilide.

A. M., 36 years old; admitted June 23, 1896. Four months ago the patient first noticed an eruption of small nodules on her arms. These nodules degenerated and formed shallow ulcers, which dried up six weeks ago. The group on the left side of the neck, at the hair-line, and the patches in the left eyebrow developed six months ago.

Present Condition.—The inner portion of the right eyebrow (Pl. 22) presents a group of hard, shiny papules ranging in size from a lentil to a pea; between the two larger ones is a tough scar surrounded by infiltration.

At the edge of the hair, in the neck, is a group of copper-colored papules, closely crowded together for the most part, resting on an infiltrated base. The center is occupied by desquamating scars continuous with the bands of infiltrated tissue, which are partly covered with scabs and appear arranged in folds; distinct, isolated nodules, somewhat larger than lentils, are disposed about the periphery. Below this is another group, the center of which consists of a purulent infiltrate, with infiltrated, desquamating edges, surrounded by fresh, pale nodules. A third patch presents a keloid appearance, the infiltration, which is similar to that in the center of the papules, being depressed in the center and at the edges. Other similar patches are found on the extremities.

In the gluteal folds and in the prolongation of the right labium majus is a group of nodules as large as the head of a pin, some of which are not eroded. Other papular eruptions, similar to those shown in the illustration, are scattered more or less profusely over the entire body; some of these show the crater-like central scar, others merely pale nodules, varying in size from the head of a pin to a lentil or a pea, resting directly on the skin.

Wherever the infiltrated areas have coalesced, the entire skin is converted into a plaque of infiltrated tissue. The crater-like scars are tough, with everted, glistening edges, and resemble keloids. The lymphatic glands, although not swollen, are hard to the touch. The patient is pale, but not anemic. She has been pregnant seven times; one child born at term and six abortions in the third or fourth month; she has menstruated regularly for the last four years. The patient has always been well, comes of a healthy family, and has no knowledge of her disease.

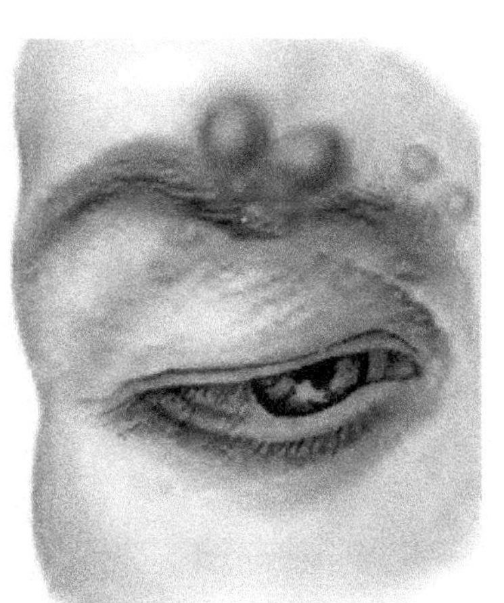

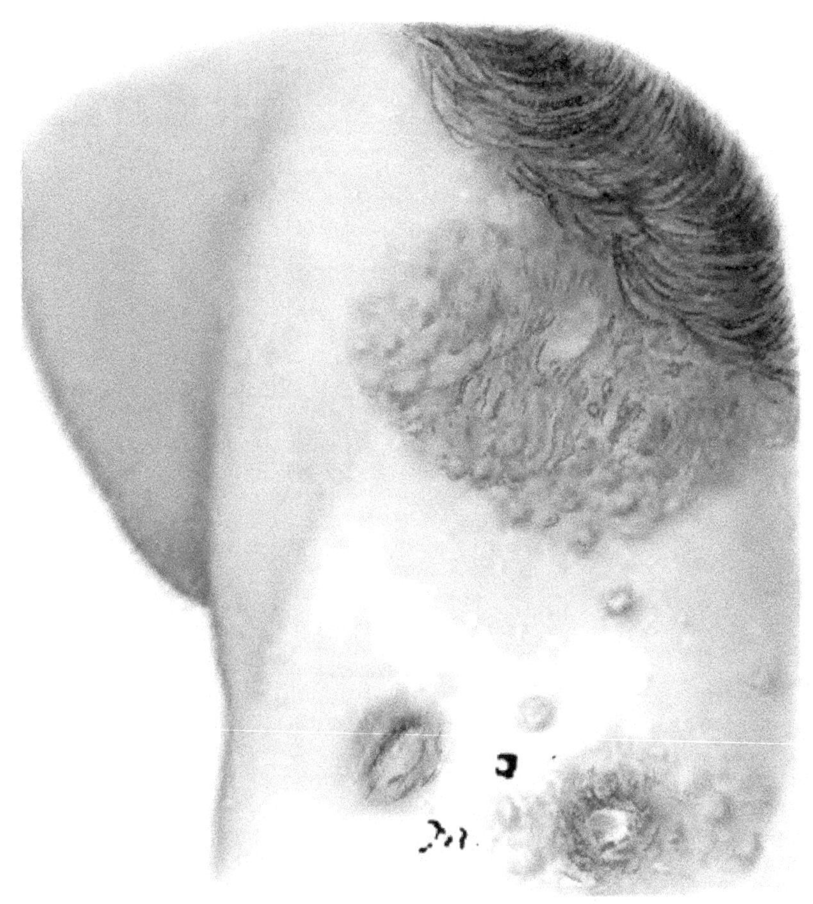

PLATES 24, 24a.

Leukoplasia of the Neck. Papules on the Genitalia.

A. B., 18 years old, servant-girl. Has never had a venereal disease. In the beginning of Dec., 1895, she began to be troubled with burning during micturition; at the same time several "pustules" developed on the outside of the labia majora, which burst after several days and healed over. There was also a painful swelling of the right inguinal glands, lasting several weeks and disappearing finally with rest in bed and compresses. In Feb., 1896, she was troubled with pain in the throat, and for two weeks was unable to swallow solid food. These symptoms improved after gargling with alum. A few days afterward an erythematous eruption appeared on the throat, on the flexor surface of both elbows, and on both legs. Since the end of March the eruption has been brown. On May 23d she came under hospital treatment; up to that time she had not consulted a physician. Last coitus six months ago; last menstruation, April 29th. Has never given birth, nor had an abortion.

Present Condition.—Eroded, edematous papules on both large and small labia, especially on the right side; inguinal glands on both sides much enlarged; at the anus the mark of an old papule. On the lower extremities a specific eruption in process of regeneration; intense leukoplasia of the neck; both tonsils enlarged and ulcerated.

Cured after twenty inunctions.

Black Plate (Pl. 24a): Front view of the same case.

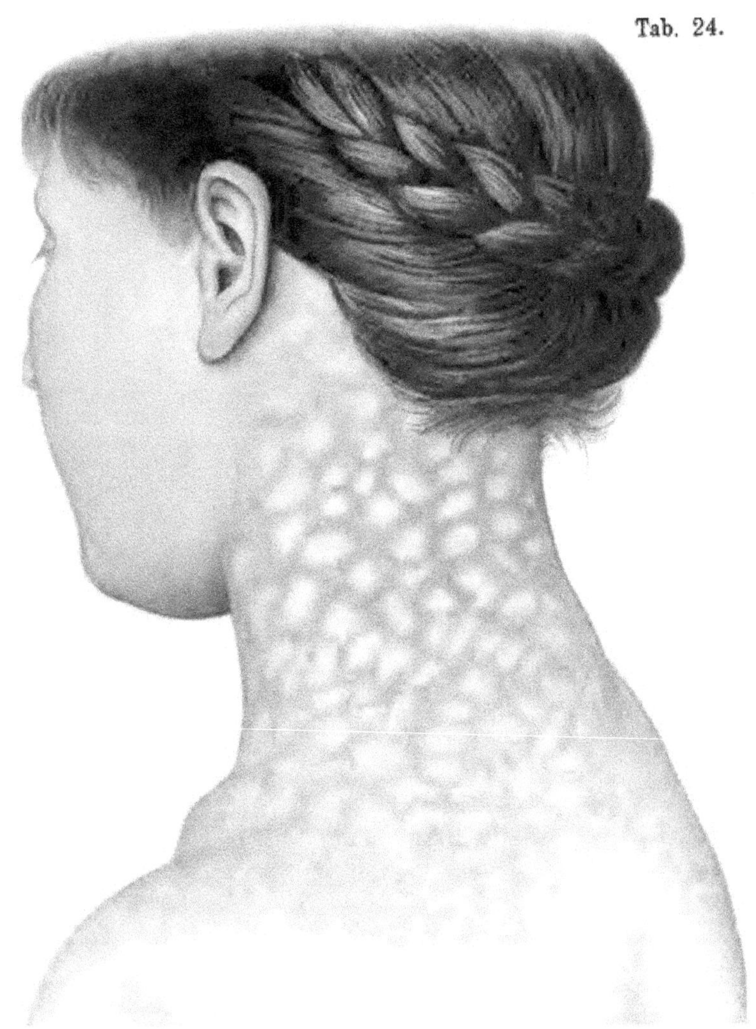

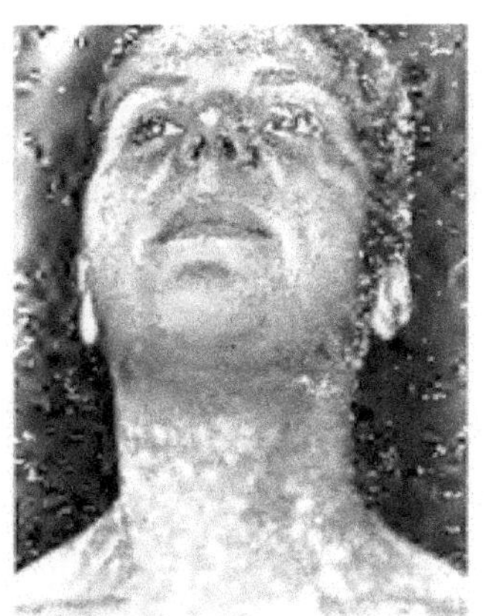

PLATE 25.

Flat, Glistening Papules on the Forehead and Face.

N. M., 26 years old, locomotive engineer; admitted Oct. 5, 1896. The forehead presents irregular patches of red and a few papules but slightly raised above the level of the skin. There is a narrow circle of red at the periphery of the papule; the center shows a brown discoloration, while the epidermis in the intermediate zone has a tense and faintly glistening appearance. On the alæ of the nose and on the chin are similar papules, less distinctly marked.

Other symptoms: a maculopapular syphilide on the trunk; moist papules on the skin of the scrotum and on the skin of the penis; on the foreskin the scar of a sclerosis.

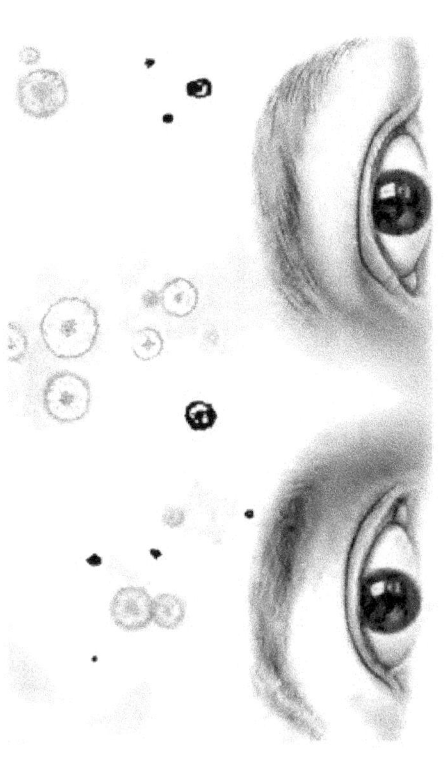

PLATE 26.

Syphilitic Alopecia Areolaris.

S. H., 25 years old, works in a brush-factory; admitted April 13, 1896. In Nov., 1895, patient had a chancre; in December he was treated in the hospital for an eruption. The present symptoms developed three weeks ago.

Present Condition.—On the forehead and on the hairy scalp are numerous pustules, some with the scabs still on them. Where a pustule has dried up and shed its scab the hair is gone completely, the base of the pustule is converted into a glistening scar, and the hair-follicle is not visible. The general growth of hair is good, but there are areas where the hair is loose, especially about the "bald spot." The rest of the body presents, in addition, a large macular syphilide on the trunk and eroded papules about the anus and genitalia. The lymph-glands in general are swollen. Additional eruptions of a papulopustular character appeared later on the hairy scalp, so that this part of the body is to be regarded as the principal seat of the syphilitic eruption.

Cured by the use of white-precipitate ointment on the head and twenty inunctions of 5 g. (ʒjss) each.

PLATE 26a.

Papules on the Hairy Scalp.

T. A., 33 years old, works in a market. Under treatment from May 26 to June 25, 1897. Patient has never had a venereal disease. He began to notice his present condition two weeks ago. Last coitus four weeks ago.

Present Condition.—Ulcerated papules on the lower surface of the penis and on the scrotum. Multiple swelling of the inguinal glands. Raised papules about the anus. Maculopapular syphilide on the trunk and extremities. Papulopustular eruption on the head, with alopecia areolaris. Cervical and axillary glands enlarged. Palmar and plantar psoriasis. The mucous membranes of the mouth and throat are not affected.

Treatment.—White-precipitate ointment. Labarraque's solution locally. Antiseptic mouth-wash. Cured after twenty-five inunctions.

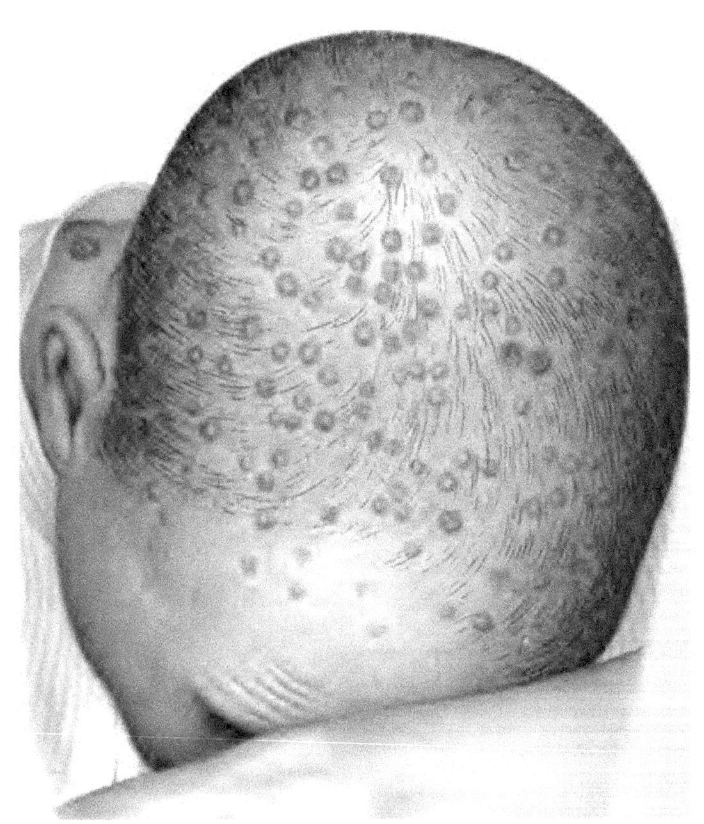

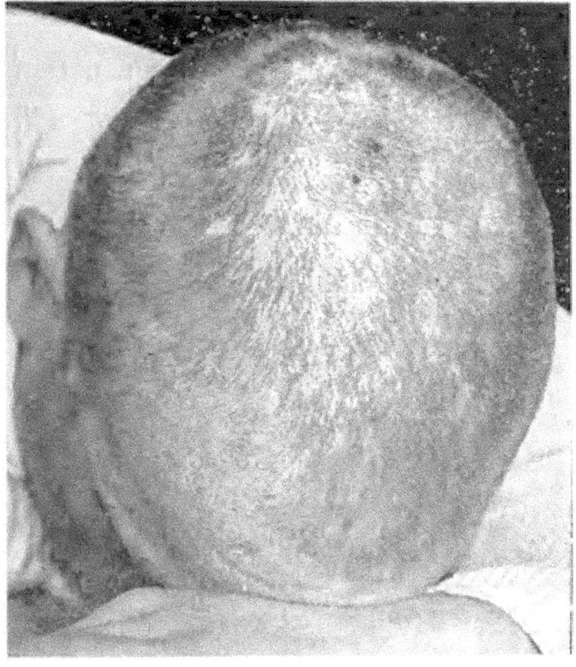

PLATE 27.

Small Pustules on the Face.

E. H., 28 years old, hostler; admitted Feb. 15, 1896. Five weeks ago an ulcer formed on the frenum. During the past week swelling and suppuration of the inguinal glands on the right side. Meanwhile a typical induration developed in the base of the ulcer, and on the trunk a scanty papular syphilide. When the inunctions were begun the eruption became more distinct and spread to the back, neck, and face. On the face the eruption takes the form of hard nodules, ranging in size from the head of a pin to a pea, and surrounded by a reddish-brown or copper-colored halo. The center of the pustule consists of a horny core, which can be easily removed, exposing the newly formed, glistening epidermis beneath. Some of the nodules are collected in groups.

Cured after twenty-five inunctions.

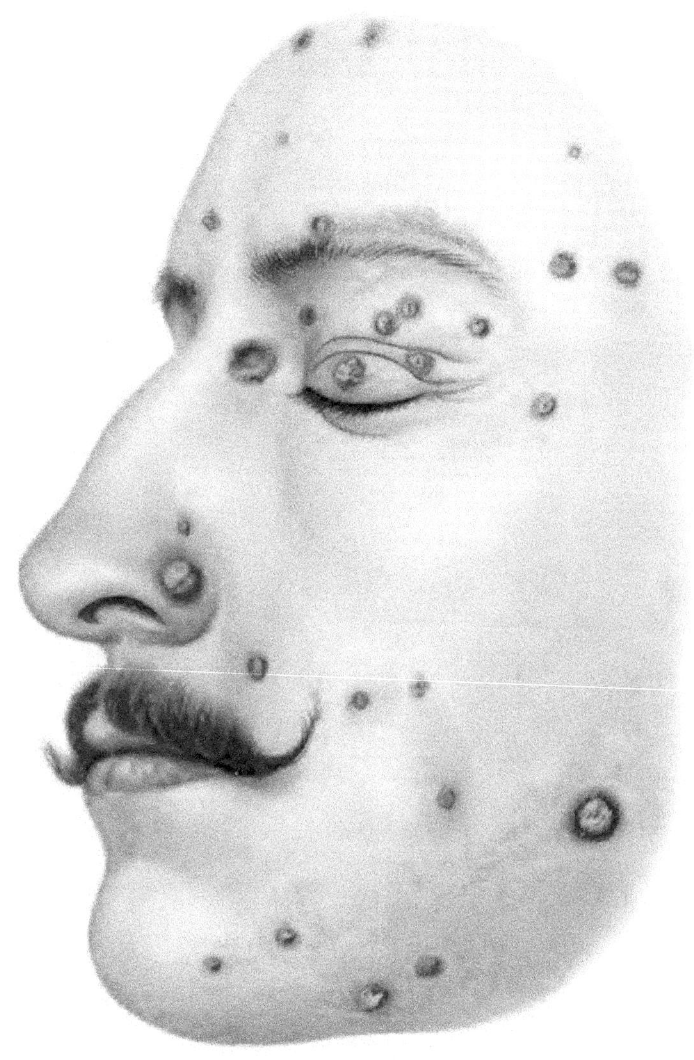

Lith.Anst.v. F Reichhold. München

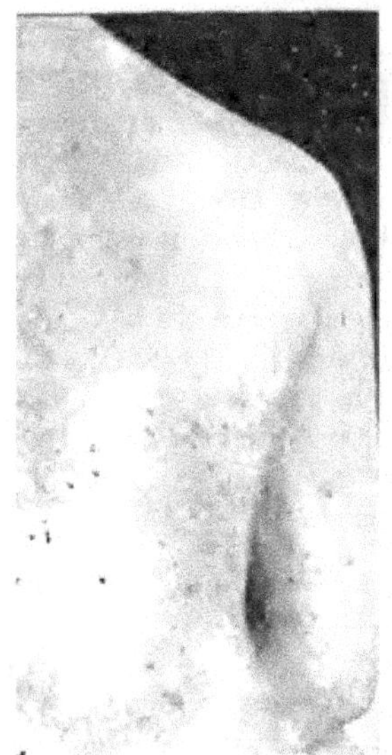

PLATES 28, 28a, 28b.

Pustular Syphilide.

P. J., 33 years old; admitted Dec. 1, 1895. Patient is sick for the first time; had his last coitus two months ago, and first noticed the eruption three weeks ago. He often suffered from sore throat when he was a child.

Present Condition.—Both tonsils are enlarged and fissured. The left presents a ragged ulcer with shreds of necrotic tissue clinging to the surface. The submaxillary glands are as large as pigeons' eggs, the middle cervical glands about the size of hazelnuts. The axillary glands are also enlarged, but the epitrochlear and inguinal glands are practically normal in size. The trunk is covered with an extensive macular syphilide. In the epigastrium are small, lichenoid papules which already show a yellow discoloration. Numerous papules and pustules, some of which are shedding their scabs, are distributed over the extensor surfaces of the upper extremities, and here and there on the thorax and back. The pustules are more numerous on the back, especially in the sacral region, and on the legs, where they are larger and run together to form eczematous pustules covered with scabs, especially about the ankles. On the upper extremities the pustules are of a light coppery hue at the periphery; those on the legs, on the other hand, are livid and of a dark coppery red.

The hairy scalp and the palms of the hands are also the seat of a papular eruption. The acneiform pustules on the legs are surrounded by extensive inflammatory areas which form an almost continuous sheet. Where the process has been going on for some time the epidermis is covered with broad, flat crusts, and comes off in large sheets wherever the pustules are closely crowded together: at first it becomes puckered over the inflamed area, then cracks, and finally loosens and comes off. The new epidermis underneath is also inflamed.

Patient was treated with hypodermatic injections of sublimate, and discharged after five weeks, cured.

Colored Plate: Part of the eczema-pustules on the left leg, seen in the black plate (Plates 28a, 28b).

Tab. 28.

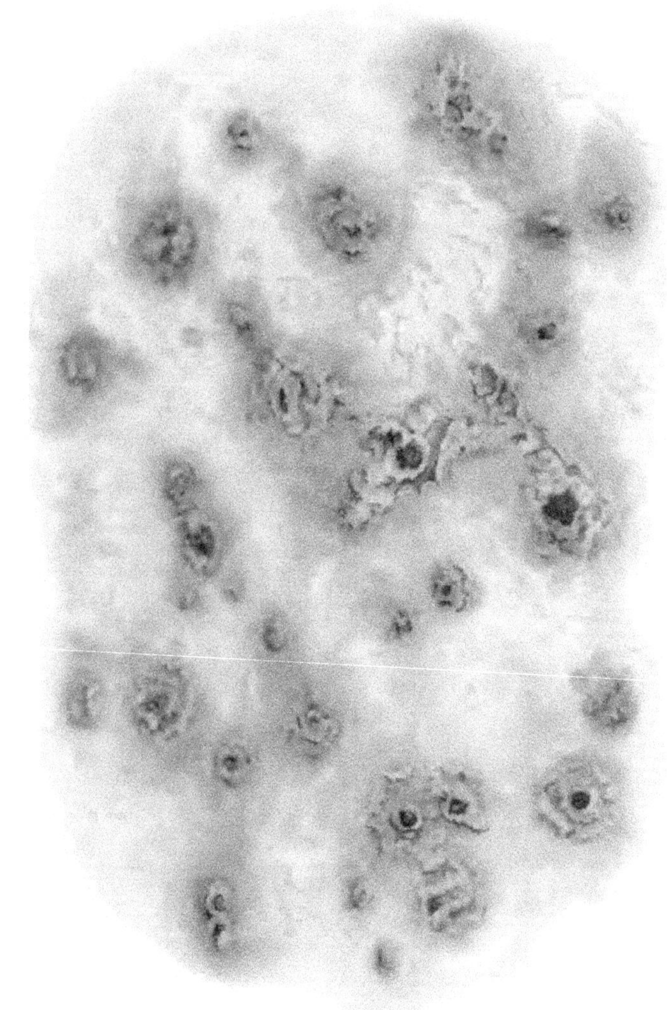

Lith. Anst. v. F. Reichhold, München

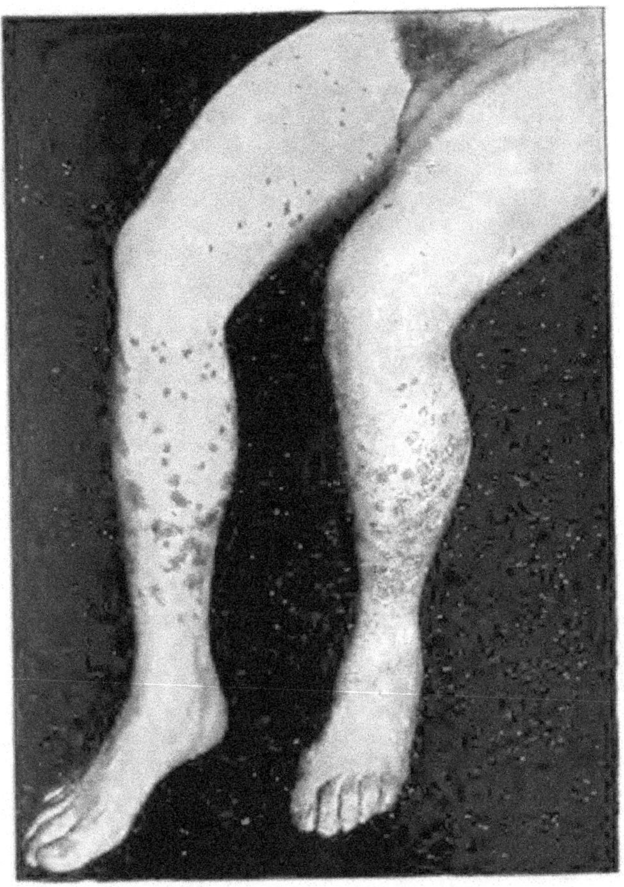

PLATES 29, 29a.

Proliferating Pustular Ulcers (Frambesia or Yaws) on Both Calves.

K. E., 18 years old; admitted Feb. 21, 1896. Patient was under treatment in this hospital last year for gonorrhea of the urethra, vagina, and canal of the cervix. A short time after she was discharged from the hospital she contracted a sclerosis on the left labium minus; was treated for six weeks and discharged after all specific symptoms, except general glandular enlargement, had disappeared. Until a week ago, the patient says, she was quite well. On that day she felt a violent itching on both legs; scratching was followed by the appearance of pustules, which were later converted into ulcers.

Present Condition.—On the right leg, below the calf, is a node as large as a dollar, composed of several smaller ones: the center is occupied by a discolored wound covered with a crust, while the periphery is made up of seven separate nodules as large as a bean, rising from 2 to 3 mm. above the level of the skin. The surface of each nodule is furrowed and, in places, destitute of epidermis so as to present fissures, while the remaining parts are covered with slightly adherent crusts of dried epidermis. The periphery of the entire node is slightly inflamed, and, like all parts of the node itself, painful on pressure. Above the large node is a fresh pustule. A similar sore, only much greater in extent, is found on the left leg (see Pl. 29a). There is typical swelling of the inguinal glands, but they are not painful. The axillary and cervical glands are also enlarged. The genitalia are flabby. Gonorrhea of the urethra and vagina. The patient says that the ulcers on the legs are painful, especially at night.

Treatment.—Inunctions. Compresses of aluminum acetate solution.

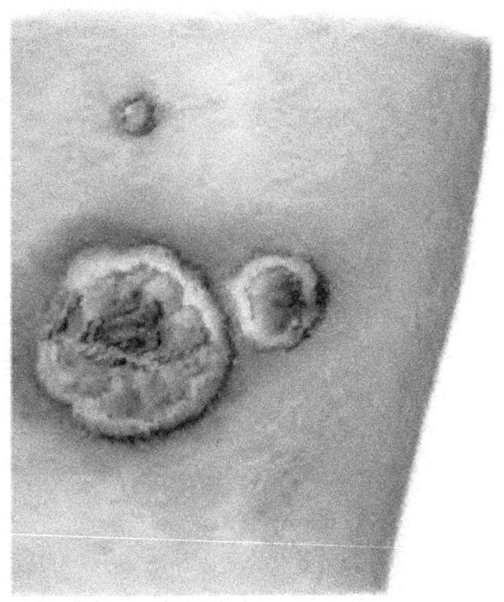

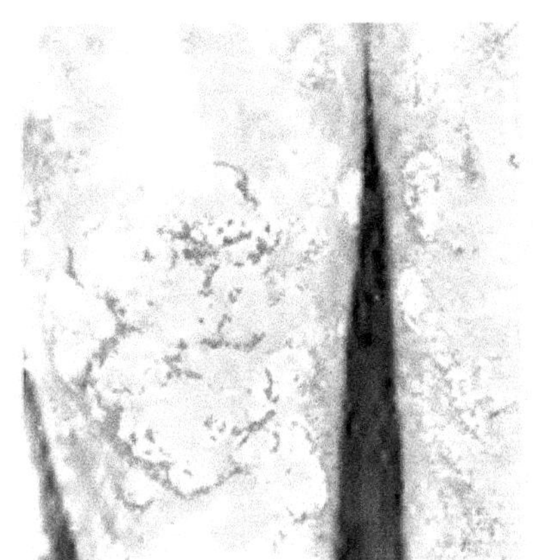

PLATE 30.

Psoriasis Syphilitica Plantaris.

R. R., 24 years old, cashier; admitted June 18, 1896. First attack. The eruption appeared five days ago. No history of a previous disease or its duration could be obtained.

Present Condition.—Numerous papules, varying in size from the head of a pin to a pea, are developing on the soles, especially in the hollow of the foot. Their peculiar reddish-brown discoloration and hard consistency indicate a horny change of the thick plantar epidermis. Numerous follicular papules are seen on both labia majora, at the commissure, and about the anus. General glandular enlargement. Pale, papular syphilide on the trunk.

Treatment. — Labarraque's solution locally. Antiseptic mouth-wash. Inunctions of 5 g. (ℨjss) ung. hydrarg.

Course.—After twenty inunctions all syphilitic symptoms disappeared. Patient was discharged July 11th, cured.

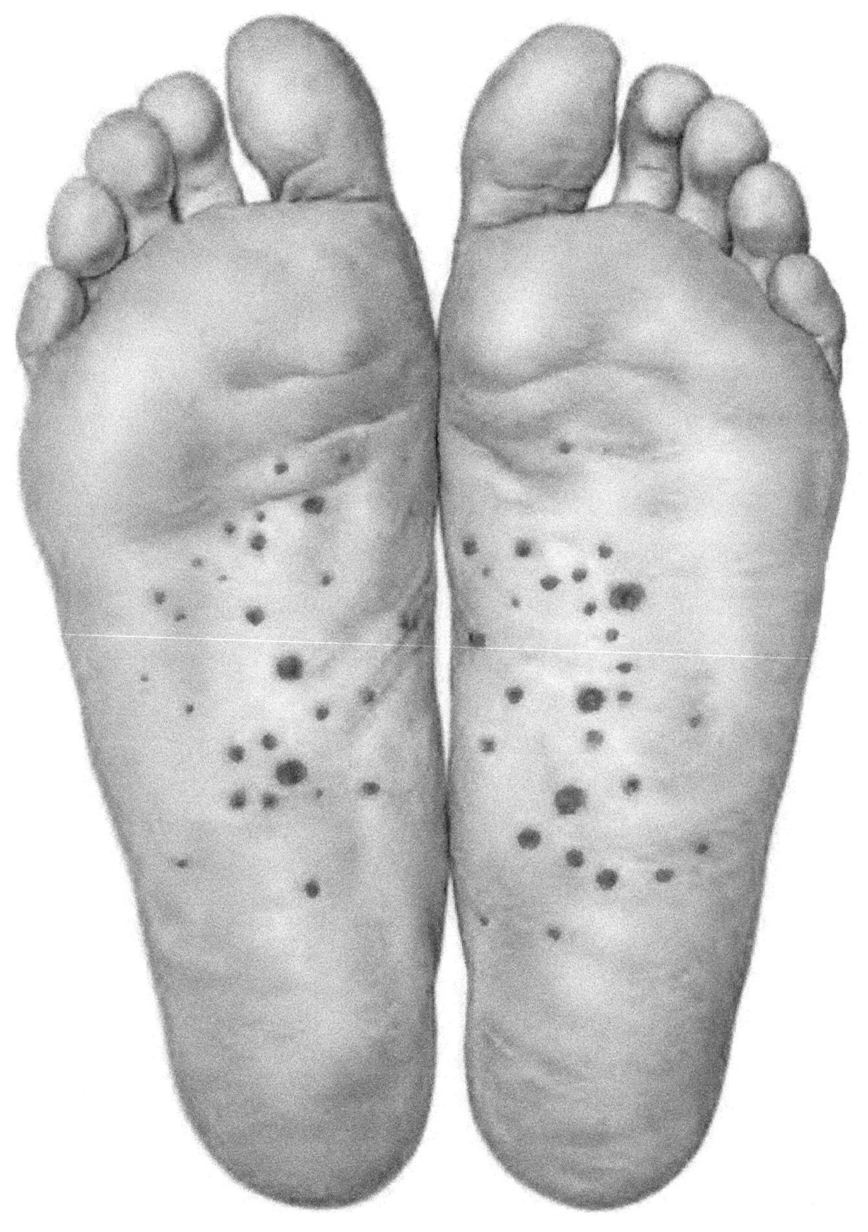

Lith.Anst. v. F. Reichhold, München

PLATE 31a.

Eroded Papules between the Toes.

T. J., 20 years old, servant-girl; admitted Nov. 25, 1896. The patient says this is the first time she is sick, and that she noticed the disease in the genitalia for the first time five weeks ago.

Present Condition.—Flat, macerated, coalescing papules on the inner surfaces of the third, fourth, and fifth toes. The toes themselves are swollen and inflamed. On the soles of both feet, papules covered with horny epidermis (psoriasis). At the edges of the labia majora and about the anus raised papules, some of which have run together. Figurate, macular syphilide on the trunk. Inguinal and cervical glands enlarged. Both tonsils are swollen and, together with the surrounding palatoglossal arches, inflamed and covered with eroded papules.

Treatment.—Sublimate baths for the feet. Applications of 5 per cent. white-precipitate ointment. Baths. Antiseptic mouth-wash. Inunctions. Cured in thirty days.

PLATE 31b.

Papules and Fissures between the Toes.

P. B., 27 years old, charwoman, married; admitted Nov. 19, 1895. The patient says she first noticed her present disease three months ago. At that time she became aware of a moist spot between the fourth and fifth toes, which she took for a corn. The other ulcers developed gradually; there was a good deal of tissue-destruction, and for the past month walking has been attended with great pain. The inflammation surrounding the ulcers has been increasing during the past three weeks; the ulcers themselves have become deeper.

Present Condition. — Degenerating papules between the first and second toes, presenting much suppuration and necrosis, with a hemorrhagic scab in the center. Between the fourth and fifth toes, which are also swollen, similar degenerating papules are seen. All the anterior portion of the foot is swollen and inflamed. The labia majora are edematous and covered with coalescing papules; on the labia minora and about the anus are isolated eroded papules.

Treatment. — Labarraque's solution (toes and genitals). Baths. White-precipitate ointment (toes). Inunctions. Cured in twenty-seven days.

a

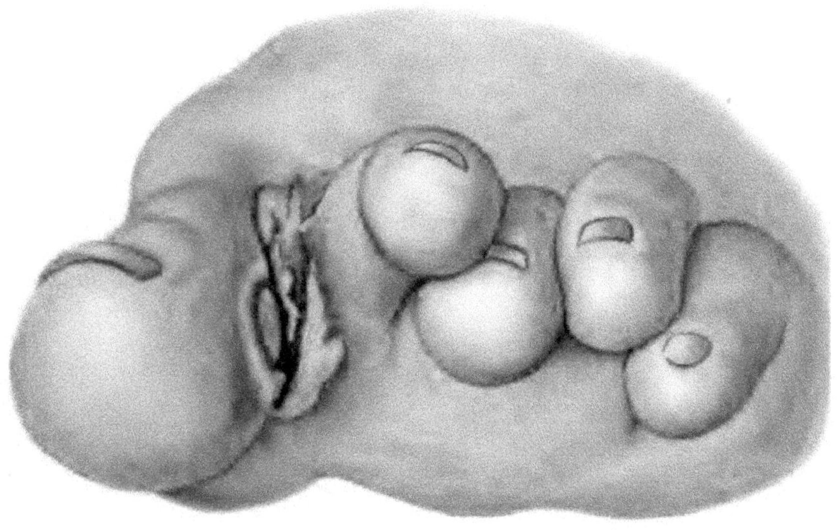

b

Lith.Anst.v. F. Reichhold, München

PLATE 32.

Syphilitic Paronychia of Both Hands.

B. J., 50 years old, laborer; admitted Oct. 5, 1896. The patient has been suffering from syphilis for the past twenty-one months, and was treated in the hospital a year ago. His present attack began a month ago.

Present Condition.—The disease has attacked the following parts with varying intensity: thumb, index and middle fingers of the right hand; index, middle, and ring fingers of the left hand; big toe of the left foot. Where the disease is mild the fingers show merely a swelling and redness of the distal phalanges and slight ulceration at the margin of the nail. Those which are more severely attacked, as the index of the right and the middle and ring fingers of the left hand, are very red, and the distal phalanx, especially the margin of the nail, swollen to the finger-tip; the nails are turned in and separated from their matrix. The latter is converted, at the margin and under the nail, into a granulating, suppurative ulcer. Papules on the buccal mucous membrane, about the anus, on the scrotum, and on both forearms. The patient complains of constant burning-pains in the tips of his fingers, which he carefully guards against injury.

Treatment.—Sublimate baths for the hands. Inunctions. Cured after twenty-five inunctions. The nails are discolored a brownish-black, brittle, and turned in at the edges.

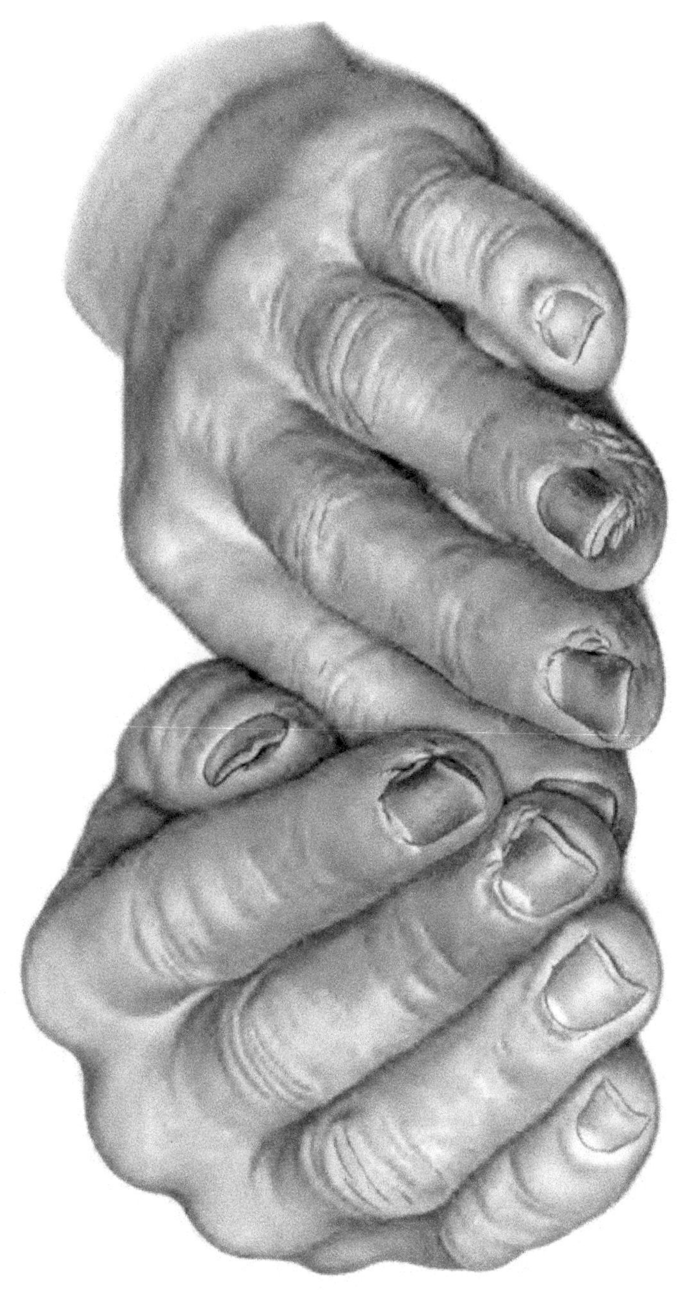

PLATE 33.

Proliferating, Eroded Papules of Diphtheritic Character.

J. M., 27 years old, coachman; admitted June 12, 1897. Patient has never had a venereal disease. Noticed his present trouble four weeks ago. Last coitus three months ago.

Present Condition.—Diphtheritic papules on the glans penis, at the edge and on the inner surface of the prepuce, on the skin of the penis, and on the scrotum. Proliferating papules on the perineum, on both thighs, and on the buttocks. Desquamating papules on the palms of both hands and the soles of both feet. Raised papules, of a livid color, covered with crusts, on the skin of the abdomen. Inguinal, axillary, and epitrochlear glands swollen.

Cured after ten inunctions and ten injections of 1 per cent. sublimate solution.

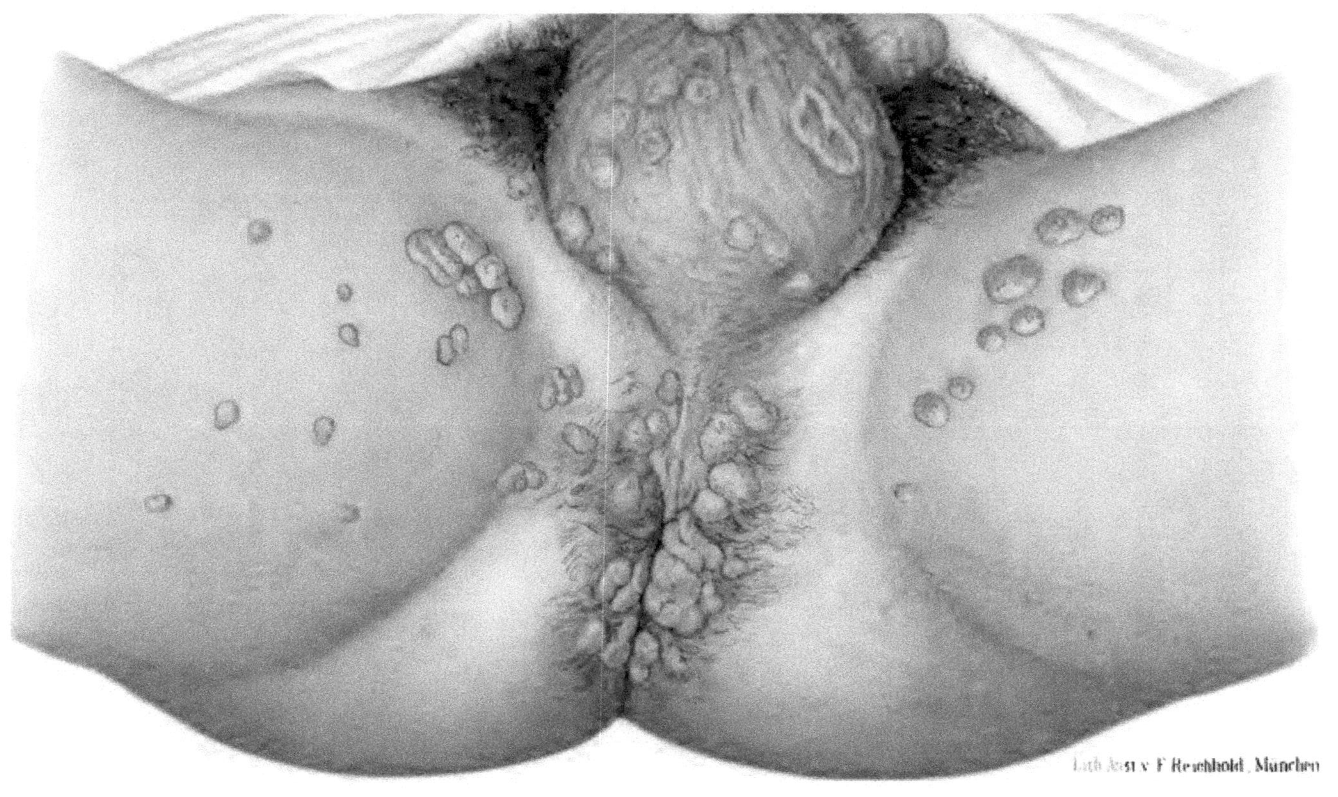

Lith Anst v. F Reichhold, München

PLATE 34.

Proliferating Papules.

T. A., 17 years old, servant-girl; admitted July 2, 1897. First venereal attack. Patient first noticed the condition of the genitals two weeks ago. Last coitus three weeks ago.

Present Condition.—At the edges of the labia majora, on the perineum, and about the anus proliferating, raised papules, some of which present necrotic decay and suppuration in the center.

Inguinal glands swollen and hard on both sides. The os displaced to the left, intact.

Cured after twenty inunctions.

PLATE 35.

Proliferating Papules on the Labia Majora, in the Genito-crural Fold, and on the Perineum as far as the Anus.

S. M., 24 years old, seamstress; admitted May 15, 1896. Proliferating, rapidly growing, inflammatory papules at the edges of the labia majora and in the anal folds; similar but smaller ones on the buttocks and the inner surfaces of the thighs. The papules are moist, but only a few present signs of degeneration and suppuration, so that this form is characterized chiefly by its inflammatory nature and rapidity of growth.

In addition, the patient is suffering from a vaginal discharge; the inguinal glands are swollen, and the mucous membrane of the isthmus of the fauces is diseased.

Treatment.—Labarraque's solution locally. Inunctions.

Tab. 35.

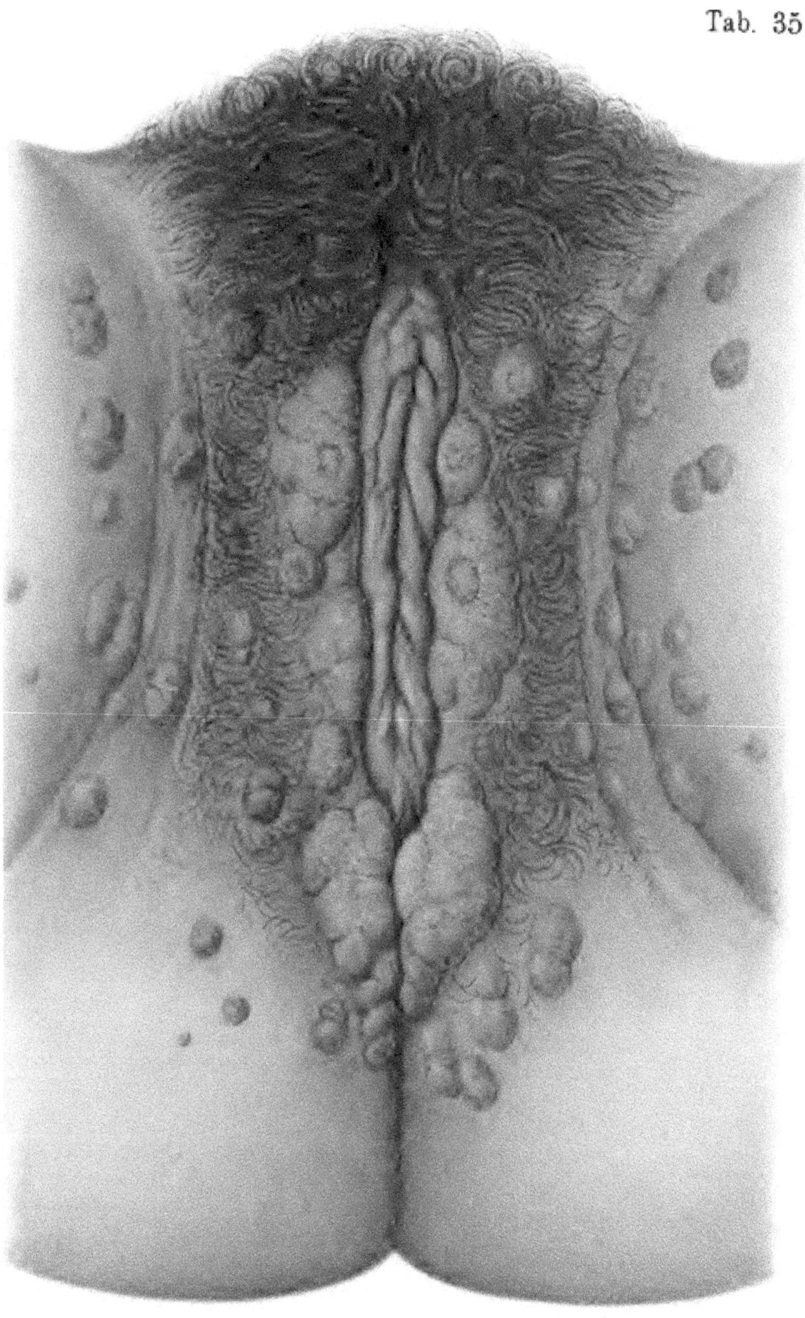

Lith. Anst. v. F. Reichhold München

PLATE 36.

Proliferating Papules on the Labia Majora, on the Perineum, and about the Anus.

G. A., 20 years old; admitted Dec. 19, 1896. Patient says she has been ill for two weeks.

Present Condition.—Raised papules, eroded at the surface, some isolated, others coalescent, on both labia majora, and extending downward over the perineum to the anus. In places the proliferations are raised as much as ½ cm. above the surrounding level, and of a hard though elastic consistence. Inguinal glands much enlarged, the remaining glands of the body only moderately so. Leukoplasia of the neck.

Cured by local application of Labarraque's solution and inunctions (twenty).

PLATE 37.

Hypertrophic Papules and Folds about the Anus.

J. T., 22 years old, laborer; admitted July 18, 1897. Patient has been treated twice for papules on the genitalia and has had altogether thirty-seven inunctions. Noticed the present trouble three weeks ago. Says he had his last coitus in Sept., 1896.

Present Condition.—Numerous livid, infiltrated folds about the anus. Close to these large, dry syphilitic proliferations, of about the size of a walnut, hard, irregularly wrinkled. In the anal fold and on the buttocks are also smaller, moist papules on a level with the skin.

The patient looks very much neglected, is covered with a tertiary macular syphilide, the glands of the body are swollen, and the mucous membrane of the throat is diseased.

Cured by thirty inunctions.

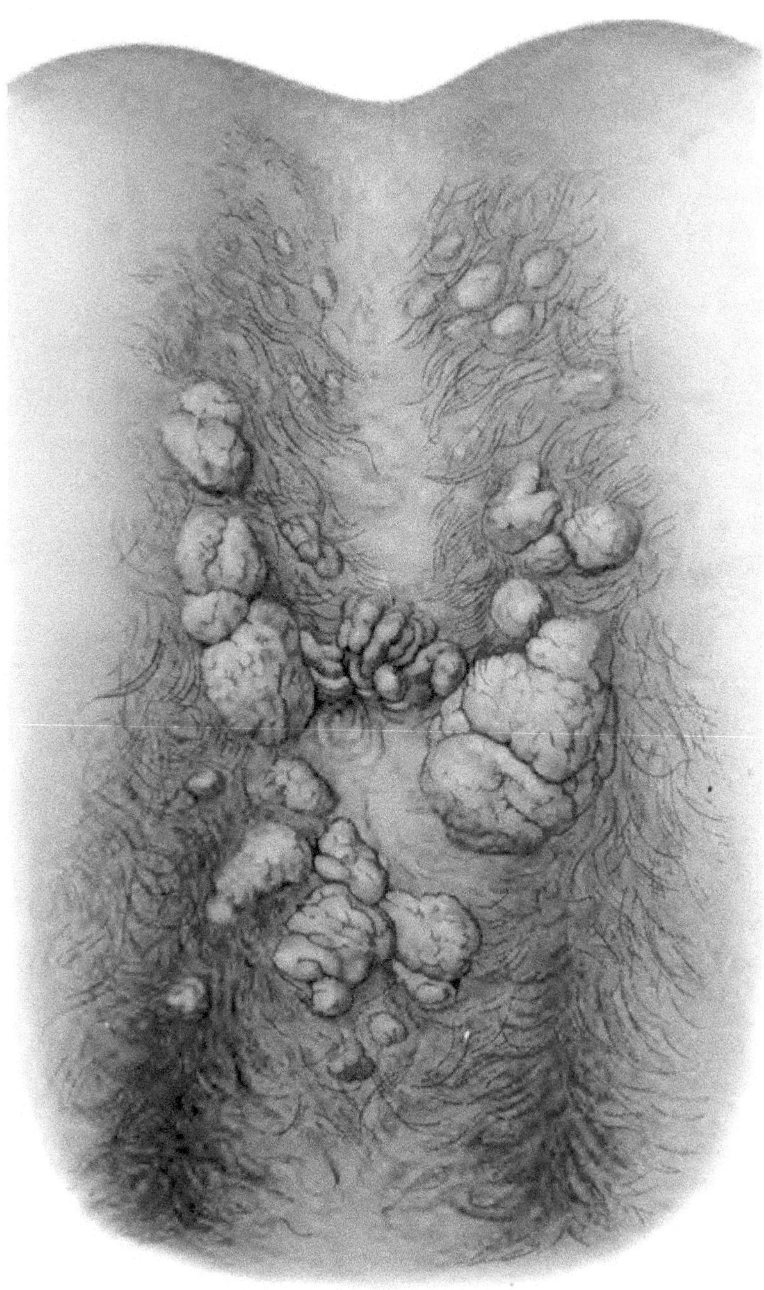

Lith Anst v. F Reichhold München

PLATE 38.

Old Annular Papules that have begun to Heal in the Center.

T. R., 17 years old, servant-girl; admitted July 29, 1897. Patient says she was treated a year ago in the hospital for a disease of the genitals which she is unable to describe in detail. Her present attack began two months ago.

Present Condition.—The labia majora, with their prolongations as far as the anus, and both groins are thickly covered with partly isolated and partly coalescent papules. Here and there, owing to regeneration of the central portion, the patches are converted into circular wreaths as large as a penny, raised above the level of the skin, with a dark-brown pigmentation in the center; or the inner margin of the wreath is degenerated at the surface, while the center is covered with whitish-gray scar-tissue. The inguinal glands are swollen. On the front of the legs and on the back and buttocks are seen a number of flat, pigmented spots as large as peas. In the right supraclavicular region is a light-brown pigmented area about as large as a dollar, with here and there, about the periphery, a few slight papular elevations. Cervical glands moderately swollen. Leukoplasia of the neck. Mucous membrane of the mouth intact.

Treatment.—Antiseptic mouth-wash. Baths. Inunctions. After twenty-five inunctions the circular groups of papules are seen to be converted into dark, reddish-brown, pigmented areas corresponding in distribution to the specific eruption. Inguinal glands shrunken.

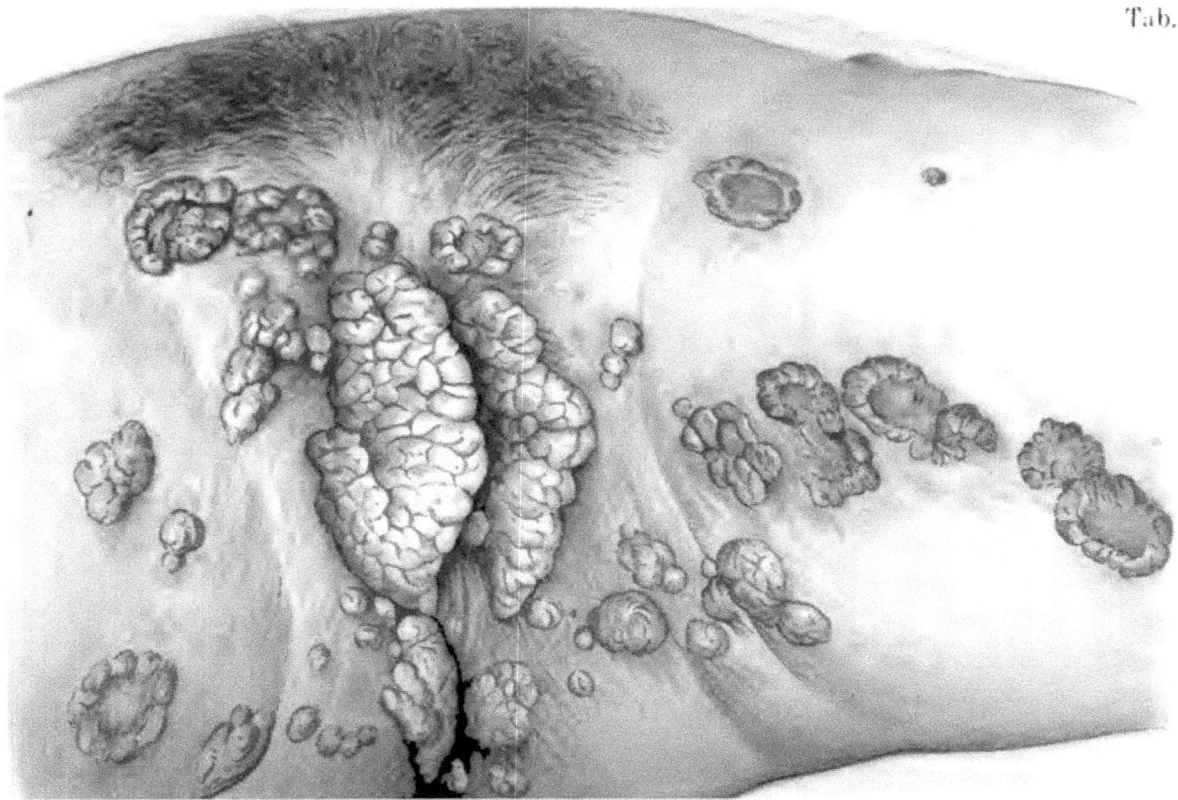

PLATE 39.

Diphtheritic Papules on the Mucous Membrane of the Os Uteri and Vagina.

M. A., 50 years old, charwoman.

The os presents fissures and contracted scars, the result of former parturitions. A number of discolored ulcers, surrounded by inflammatory tissue, are seen; two on the anterior, one on the posterior lip, and several, partly coalescent, on the posterior wall of the vagina. The patient is not aware of the ulcers in the vagina and os uteri.

Isolated, moist papules are seen on the labia, in both inguinal regions, on the perineum, and on the inner surfaces of both thighs. On the trunk and neck a pustular syphilide, mingled with papules. The inguinal as well as all the other glands of the body are enlarged.

Patient has passed the climacteric; she says she has noticed a discharge and the "ulcers" for the last two weeks. She has given birth to seven children, the last one eighteen years ago.

Sublimate was applied locally to the genitals, and the patient was subjected to nine inunctions and twenty injections, as a general treatment. After being under treatment eighty-seven days she was discharged cured.

This case dates back to the time when the author was assistant in Siegmund's clinic, 1879–80.

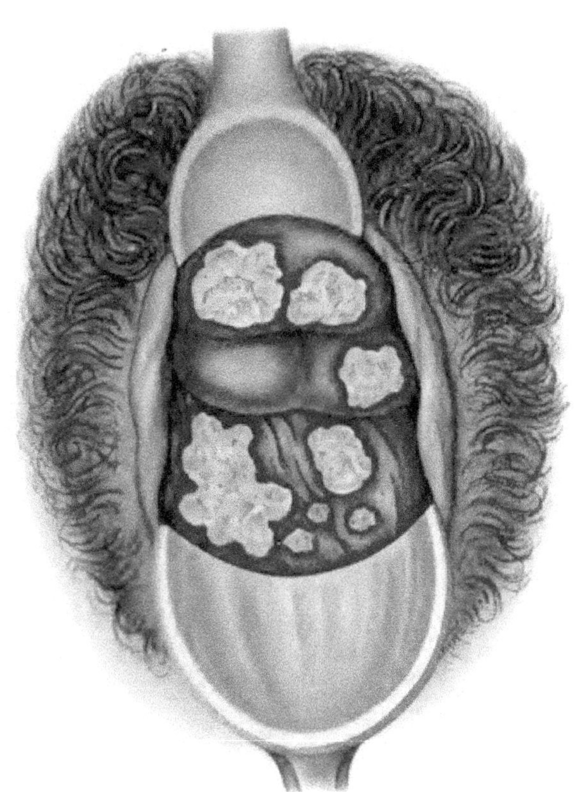

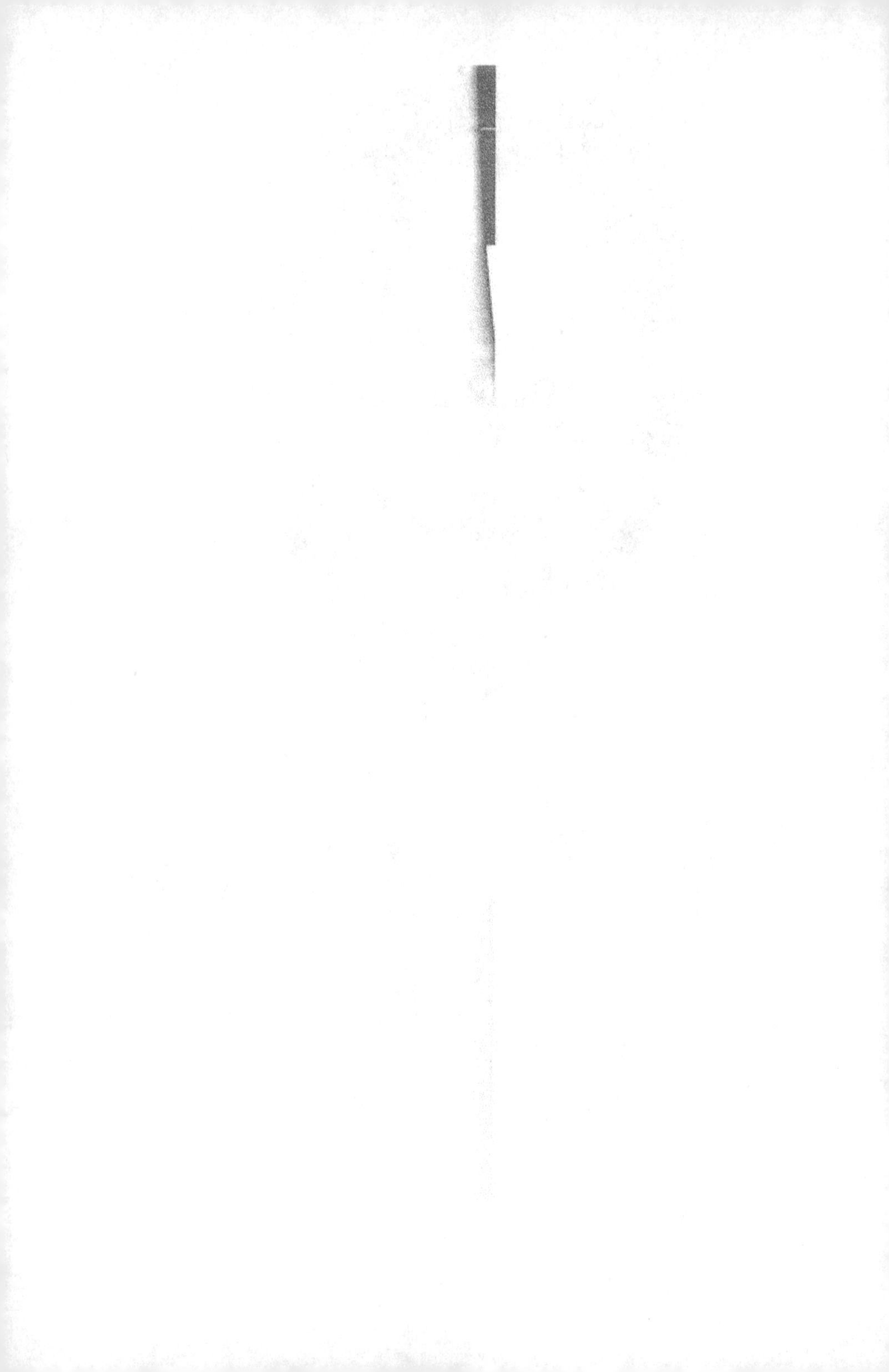

PLATE 40.

Diphtheritic Papules on the Mucous Membrane of the Upper Lip and Left Side of the Mouth.

T. A., 33 years old, tanner; admitted Nov. 12, 1896. The patient was treated a year ago for syphilis. The ulcers on the scrotum and in the mouth made their appearance four weeks ago.

Present Condition.—In the mucous membrane of the upper lip and of the cheek near the left angle of the mouth, and on the tonsils, are several discolored papules, with deep, ulcerated centers. Remnants of papules on the palms of the hands. Partly healed papules on the penis, scrotum, and buttocks. On the trunk and extremities brown pigmented papules in process of regeneration. General glandular enlargement.

Treatment.—Sublimate mouth-wash. Labarraque's solution externally. Cured after twenty-five inunctions.

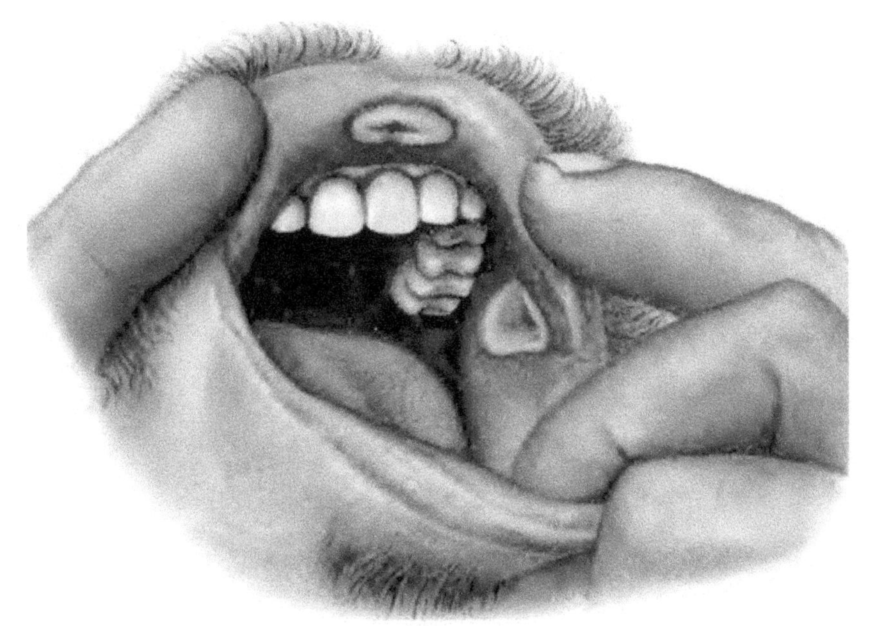

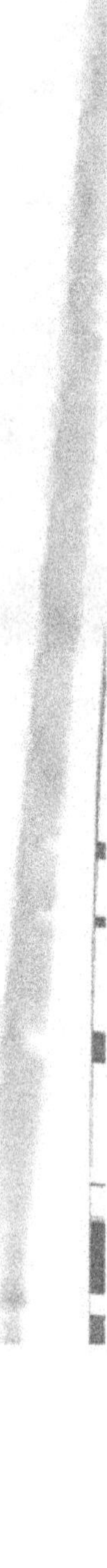

PLATE 41a.

Infiltration and Superficial Necrosis of the Mucosa and Submucosa of the Upper Lip.

K. T., 70 years old, workman in a gas-factory; admitted Aug. 11, 1896. Patient noticed the swelling on the upper lip for the first time in May of last year. He says he was never sick before; denies syphilitic infection.

Present Condition.—About the middle of the upper lip an elliptical infiltration about as large as a half dollar, the long axis corresponding with that of the lip. At the left extremity a fissure about 5 mm. wide and ½ cm. long. The submaxillary glands can be felt on both sides, but not the parotid lymphatic glands. The lymph-glands of the rest of the body but little affected.

Treatment.—Inunction. Cured after twenty applications.

PLATE 41b.

Ulcerating Papules and Incipient Leukoplasia of the Tongue.

P. P., 49 years old. Has been treated as an out-patient. The patient says that four years ago she noticed fiery-red, isolated nodules on the tongue for the first time. Various remedies were tried, among them cauterization (with lunar caustic), which caused the nodules to disappear for a time, but they always recurred. A year ago they again appeared, and the patient underwent twenty inunctions, whereupon the eruption subsided. Two months ago the nodules began to develop again, and with them whitish, coalescent ulcers.

Present Condition.—The tongue is only slightly swollen; at the back the papillæ are still intact; the front is smooth and covered for the most part with a cloudy, whitish layer of epithelium. A discolored, slightly raised ulcer extends across the tongue and along both margins, while a similar ulcer, as large as a pea, occupies the tip of the tongue a little to the left of the center. The ulcers are slightly raised above the surface and surrounded by a sharply defined inflammatory border.

The submaxillary glands are hard and moderately swollen. Painful mastication.

After the patient had been treated for eight days, scar-formation began in the middle of the ulcer, which was finally converted into a whitish, epithelial hyperplasia.

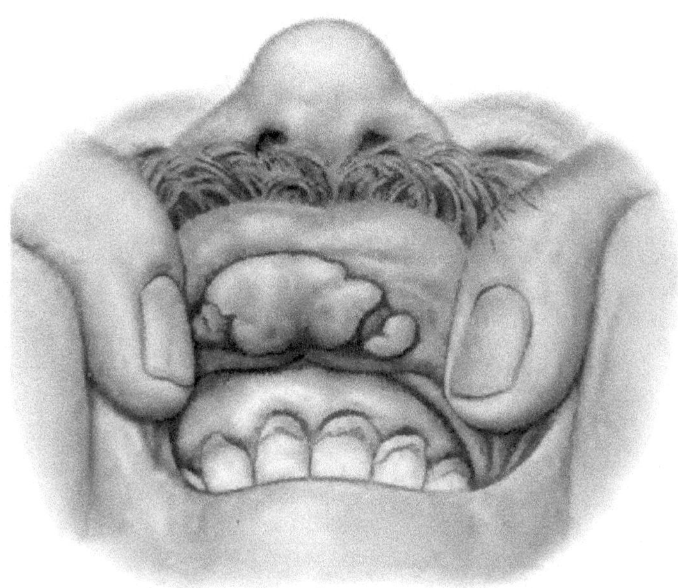

a

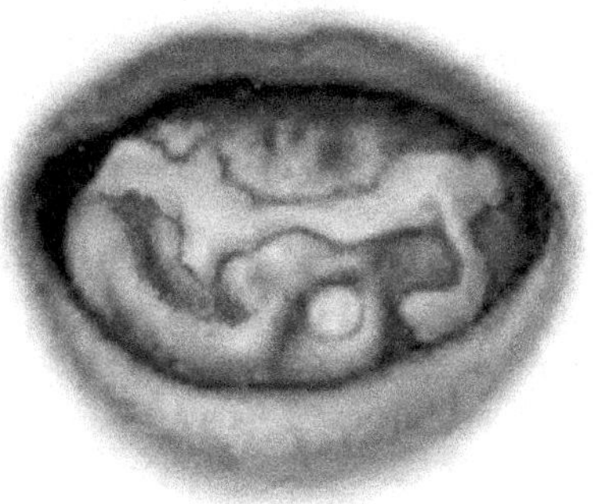

b

PLATE 42a.

Elevated, Coalescent Papules on the Hard Palate.

R. S., 21 years old, prostitute; admitted Nov. 16, 1896. The patient was first infected in 1893, and has since been treated nine times for syphilis. Most of the relapses consisted in papular eruptions on the genitals. The present attack first attracted the patient's notice two weeks ago.

On the hard palate, stretching from the fossa behind the incisors to the soft palate, is a coalescent group of mulberry-like proliferations of hard, yet elastic consistence, somewhat lighter in color than the slightly inflamed mucous membrane of the surrounding parts. The edges of the soft palate and uvula are slightly thickened and distorted as the result of a former attack of the disease, which even now betrays itself by an infiltration on the edge of the soft palate and uvula. The vibrations of the pillars of the fauces during phonation are sluggish and irregular. Concomitant symptoms are found in flat, glistening papules, as large as a bean, on the labia majora, and in a general glandular enlargement.

Treatment.—Inunctions. The specific infiltrations disappeared, the proliferations on the hard palate subsided, and the mobility of the pillars became almost normal.

PLATE 42b.

Leukoplasia (Psoriasis) Linguæ.

C. J., 49 years old. Under treatment for emphysema and pulmonary catarrh in Ward No. 12.

The patient has had various diseases. In 1872 or 1873 he acquired a hard chancre, which was followed by eruptions on the skin and sores in the mouth. With the exception of local remedies and river-baths the patient did not undergo any treatment for his disease. Lunar caustic, gargles, and precipitate ointments were the local remedies he employed.

The patient used to be a heavy smoker; when he worked on a freight train he used to smoke both cigars and pipe day and night. In 1891 he noticed for the first time whitish

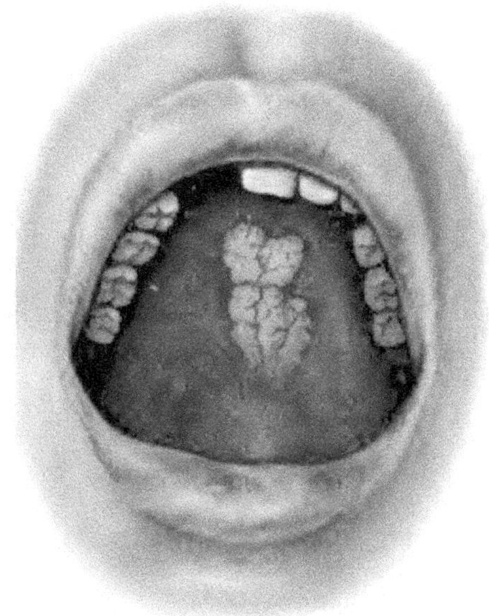

a

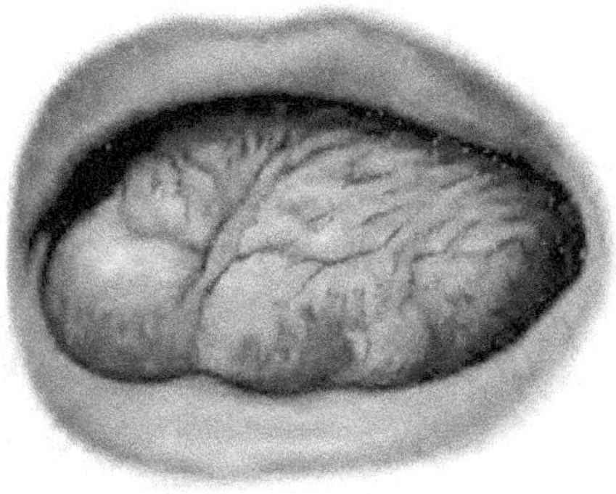

b

vesicles on the tongue, which bled when they were opened with a pin. The present condition of the tongue the patient says he has noticed for the last eighteen months. He is thin, but not cachectic.

Present Condition.—The tongue is not perceptibly swollen; but the patient can only protrude it a little and with difficulty. The surface is white, moderately thickened, and divided into irregular islands by shallow grooves. These grooves do not appear to be due to contracting scars, but rather to correspond to the normal furrows in the tongue. On the other hand, the islands appear slightly raised, owing to the thickening of the epithelium and the moderate inflammation which preceded their formation, and which the patient described as blisters. The tongue does not feel hard, and in its present condition is not painful. All delicate tactile sensibility is lost.

The chewing of highly seasoned food or sharp pieces of bread is apt to produce fissures, which, however, heal of their own accord in a few days. The epithelium of the buccal mucous membrane opposite the alveolar border is also somewhat cloudy, but not as thick as that of the tongue.

Submaxillary glands are not swollen. No demonstrable syphilitic symptoms.

PLATE 43a.

Condylomatous Iritis.

L. P., 23 years old, footman; admitted Nov. 30, 1896. The patient complains of pains in the right temporal region, and tearing pains in the right eye for the past five days. The lids of the diseased eye were adherent; lachrymal secretion very abundant. Syphilis denied.

Present Condition.—Ciliary congestion of the right eye. Cornea and aqueous chamber normal; pupil dilates in the form of a kidney upon application of atropin, owing to a sharp synechia at the external inferior portion. From the inferior pole of the external quadrant a reddish tumor as large as a hemp-seed projects into the pupil. On the raphé of the penis, about the middle of the pendulous portion, is a moderately infiltrated, pigmented scar of a livid coppery hue, about as large as a bean. Multiple, indolent swelling of inguinal, axillary, and cervical glands. The trunk is covered with a diffuse syphilide consisting of small pustules.

Palms of the hands, soles of the feet, and buccal mucous membrane intact.

Subconjunctival injections of sublimate. Inunctions. Cured.

PLATE 43b.

Gummatous Tarsitis of the Left Eye. Trachoma.

E. H., 24 years old; admitted Nov. 14, 1895. The woman has been suffering from trachoma for several years. Three years ago she contracted syphilis and had a rash on the entire body, for which she underwent an inunction cure. For the past week she has felt a tumor under the left upper eyelid.

Present Condition.—The patient is pale and delicately built; lymph-glands generally are enlarged. Front and back of the neck covered by a typical leukoplasia. Both tonsils are enlarged and fissured.

Condition of left eye: the conjunctiva of the lower lid presents various alterations due to trachoma. A tumor about as large as an almond can be felt through the upper lid. Tarsal conjunctiva velvety and deeply injected. The conjunctiva over the convex border of the tarsus and the intermediate portion is converted into a brawny wheal, which merges internally into the slightly infiltrated semilunar fold. About the center of the wheal is a shallow ulcer, about as large as a pea, with grayish-white, discolored floor and indurated margin.

Treatment.—Inunctions. Potassium iodid internally. Cured after thirty inunctions.

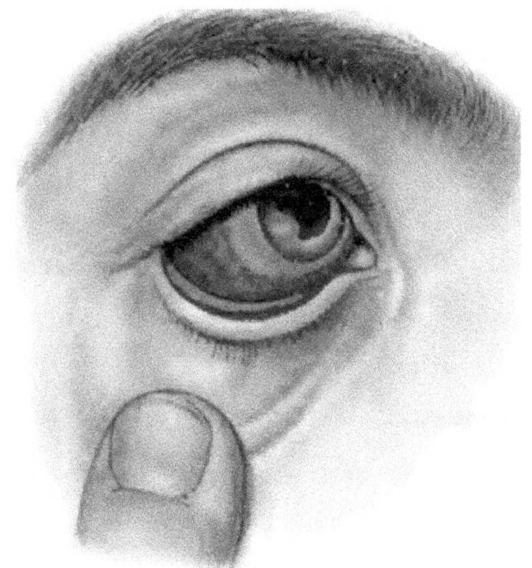

a

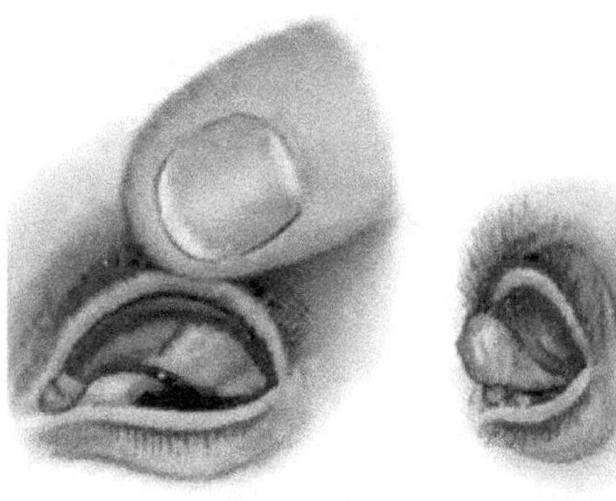

b

PLATE 44, 44a, 45.

Syphilitic Frambesia (Yaws). *Syphilis Præcox.*

J. R., 25 years old, prostitute; admitted April 6, 1896. In April, 1895, the patient was treated for a soft chancre on the genitals. In October, 1895, she acquired a hard chancre on the right labium majus; a short time afterward an eruption appeared. The patient was subjected to thirty-five inunctions. Present attack began four weeks ago.

Present Condition.—On the hairy scalp (Plates 44, 44a, 45) are several papillomatous, warty excrescences as large as a half dollàr, covered with scales and crusts. Serpiginous ulcerations on the cartilage and left ala of the nose, on the right upper arm, below the left mamma, and on the back; here and there on the trunk a few papular infiltrations.

Fig. 45 represents the same case: Mulberry-like proliferations on the hairy scalp after the crusts have fallen off.

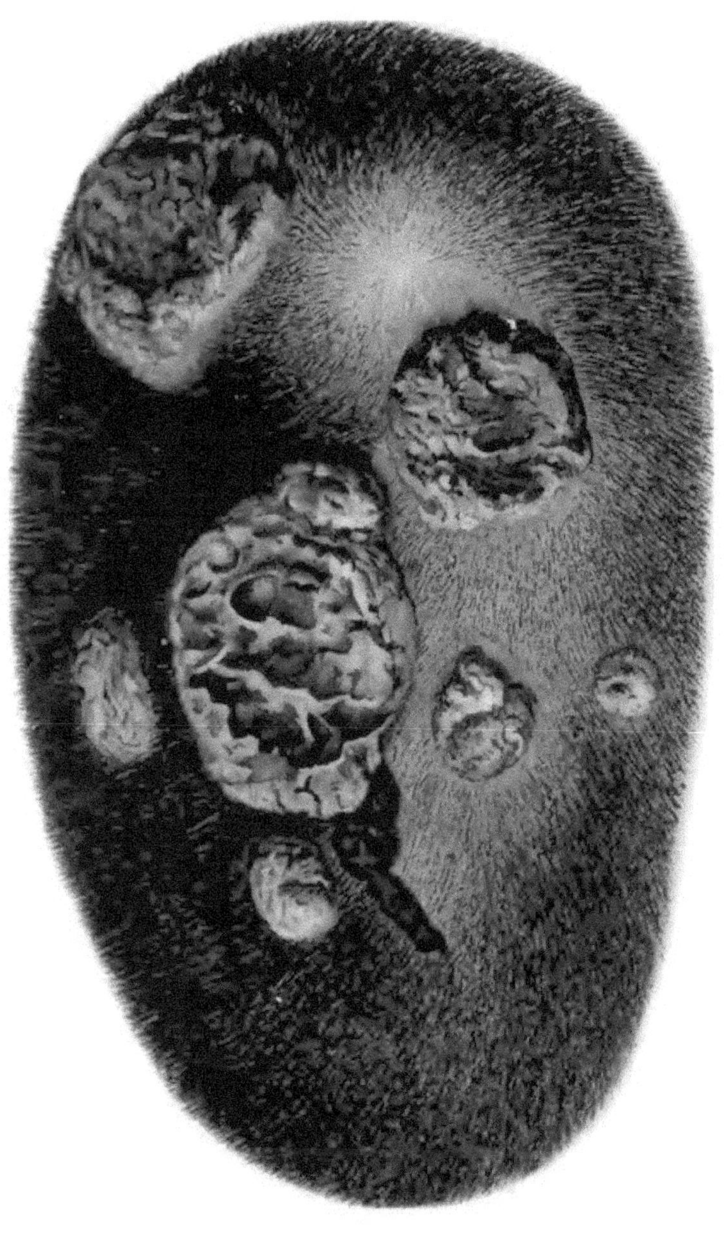

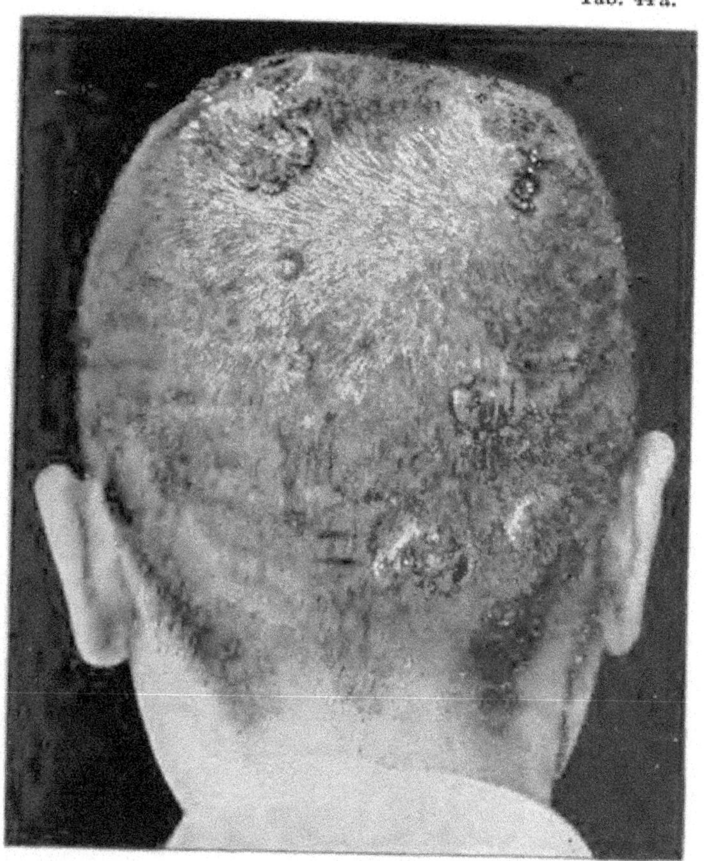

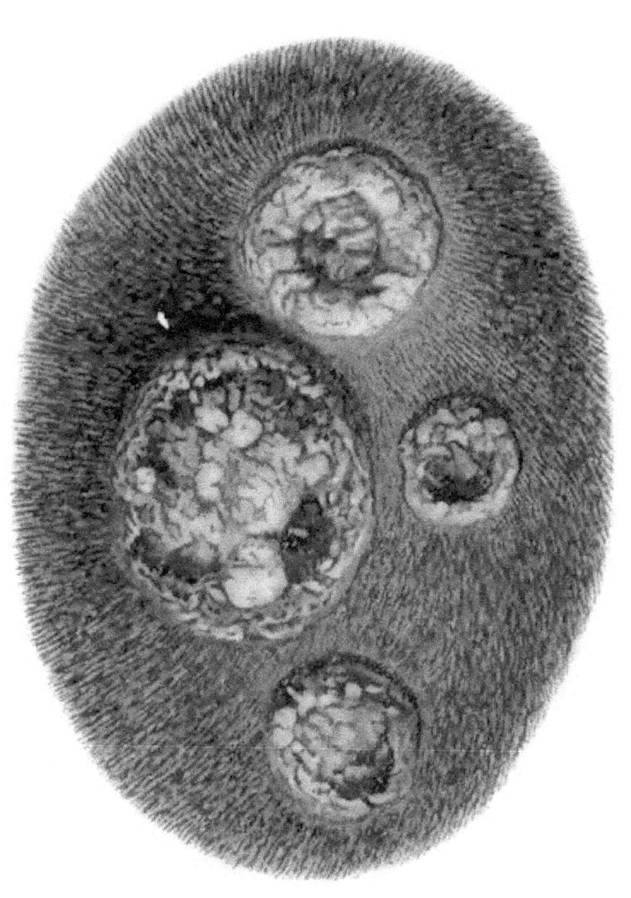

PLATES 46, 46a, 47.

Gummatous Ulcers on the Labia Majora, the Posterior Commissure, the Right Labium Minus, and the Vagina.

W. A., 26 years old, servant-girl; admitted June 15, 1896. She has been syphilitic since 1890, and is now undergoing treatment for the third time in this hospital; the first attack occurred in 1890, the second in the latter part of 1892. The present attack began only a few weeks ago, so that the patient has been free from any noticeable syphilitic symptoms for three and a half years.

Both labia majora, the clitoris, and the labia minora are hypertrophied; their consistence not perceptibly increased. At the margin of the left labium majus are three circular ulcers as large as a bean; at the margin of the right (Plates 46, 46a), two about twice as large, and on the posterior commissure two still larger ones, separated by a narrow bridge of tissue. These ulcers have sharply defined edges; the base is irregularly degenerated, and they are for the most part covered with pus. Similar, smaller ulcers, to the number of about eleven are found on the external surface of the right labium minus, in the vestibule, and about the middle of the vaginal canal. They have the same degenerated base, but are neither as large nor as deep as the others (Pl. 47).

At the right angle of the os is a discolored ulcer, larger than a bean. The inguinal glands on both sides are palpable and spindle-shaped. Pigmented scars on the legs, and here and there on the trunk.

Treatment.—The ulcers were sprinkled with iodoform powder. Twenty-four grains of potassium iodid per diem. After a month most of the ulcers had healed. The labia were still enlarged. The patient had improved a good deal, and an inunction treatment was ordered. Cured after twenty inunctions.

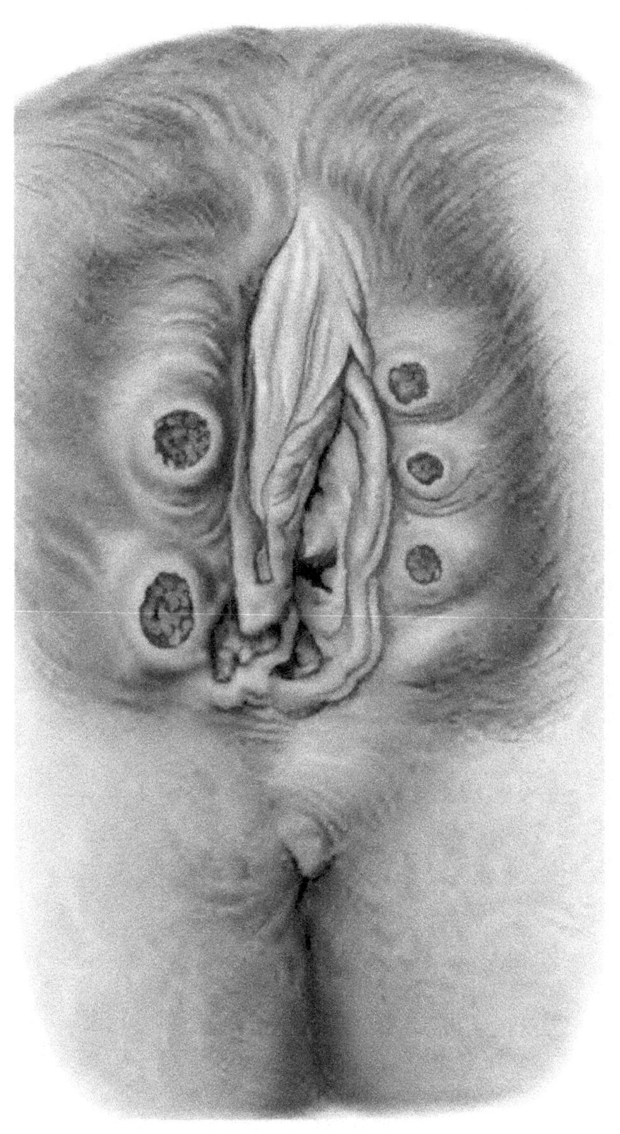

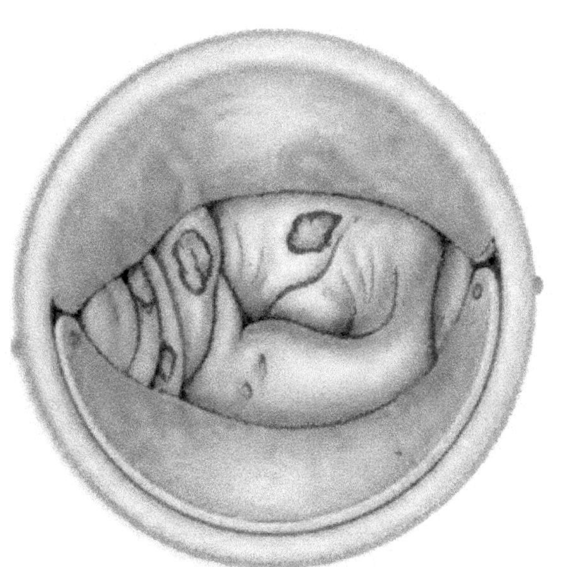

PLATE 48a.

Gummatous Ulcer of the Left Nipple.

H. R., 26 years old, drummer; admitted Oct. 9, 1896. Patient has been suffering from syphilis for six months, and has been almost constantly under mercurial treatment during that time; in spite of that, however, the entire body became covered with ulcers, to the number of seventy-four. He is also suffering from an affection of the left elbow-joint.

Present Condition.—The left nipple is replaced by an ulcerating sore; the areola is swollen and converted into a deep infiltration. On the hairy scalp are several pustular sores covered with crusts. In the right nostril a fissure, with infiltration of the base and ala of the nose. On the root of the penis a broad scar from the sclerosis. Numerous partly healed sores and pigmented and desquamating scars, the remains of the above-mentioned sores, are scattered over the entire body. The patient is emaciated and anemic; he complains of lassitude and headache.

Treatment.—Decoctum Zittmann.[1] Inunctions.

[1] Decoctum sarsaparillæ compositum (sarsaparilla, senna, glycyrrhiza, fennel and anise, with calomel, cinnabar, and alum).

PLATE 48b.

Gumma of the Breast.

C. R., 41 years old, charwoman; admitted May 28, 1897. For the past year the patient has noticed a tumor in the left breast, which gradually grew larger and began to ulcerate last fall. She says she has always been well otherwise. She has borne four living children; has never nursed; never had an abortion.

Present Condition.—In the outer half of the left mammary gland are a number of old and recent scars, which palpation shows to be due to infiltrated bands of tissue, radiating over the gland almost as far as the left margin. About the anus, on the perineum, and on the posterior surface of the labia majora the marks of old papules, surrounded by a red halo. Multiple swelling of the inguinal glands on both sides, and also of the epitrochlear and axillary glands. Mouth and throat intact.

Treatment.—Potassium iodid, 24 grains per diem. Gray plaster.

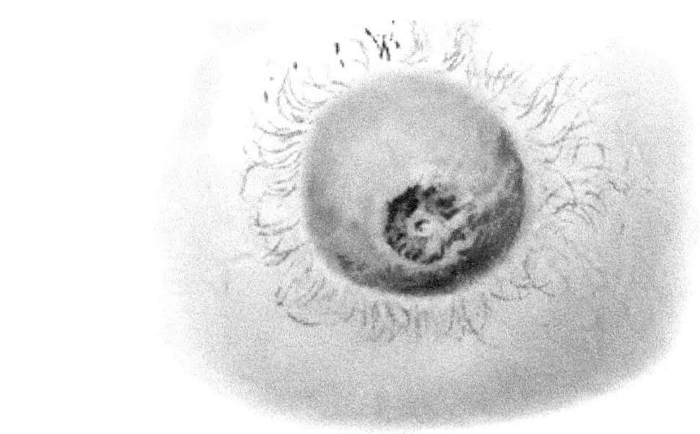

a

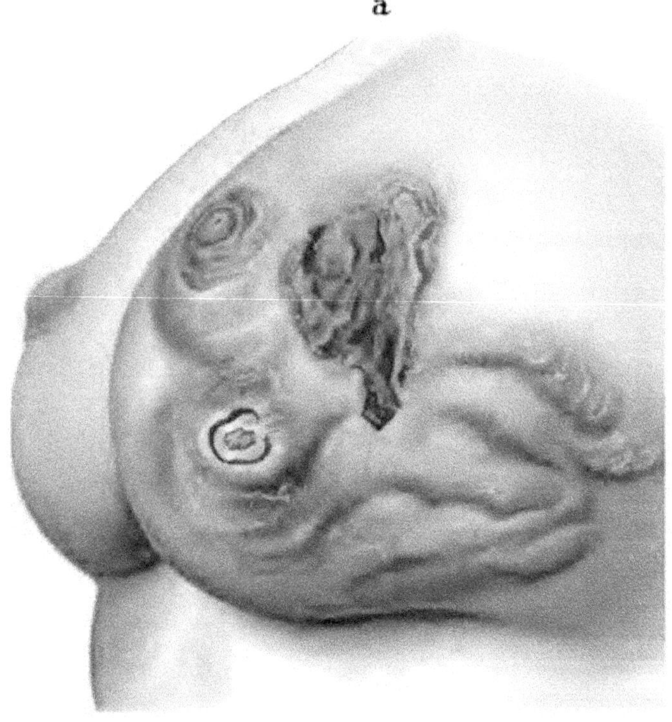

PLATE 49.

Syphilitic Rupia.

R. M., 46 years old, waiter's wife; admitted June 12, 1896. With the exception of varicella in her twenty-second year, the patient says she has never had any disease.

Five or six years ago the patient suffered from violent headache; two years ago she had an ulcer on her leg. Patient had borne twice, in her twenty-fourth and in her twenty-seventh year (out of wedlock); the children lived for some time; no abortion. Venereal disease denied. Patient is a heavy drinker.

Ulcers first appeared on the arms four months ago.

Present Condition.—Patient is very much emaciated. On the left arm, above the elbow, an oval group of infiltrations, covered for the most part with rupia-like scabs. The scars are only skin-deep. About the periphery, especially at the upper part, some desquamation of the epidermis.

In two places recent, superficial infiltrations are seen, over which the skin is raised, forming two cloudy blisters.

On the right arm, extending below the elbow, a semicircular group of similar sores, some of them already converted into scars, some still covered with rupia-like scabs, and some quite recent, resembling blisters. General glandular enlargement.

Whitish scars are seen in the left groin and at the edge of both labia majora.

Internal organs normal, with the exception of an old process at the apex of the right lung.

After the scabs had fallen off, round and oval ulcers appeared in the affected areas, penetrating the skin and covered with pus.

Scar-tissue began to form after the ulcers had been treated with red-precipitate ointment for two weeks, and iodid of iron had been given internally. With the exception of a slight redness at the site of the ulcers nothing abnormal in the skin.

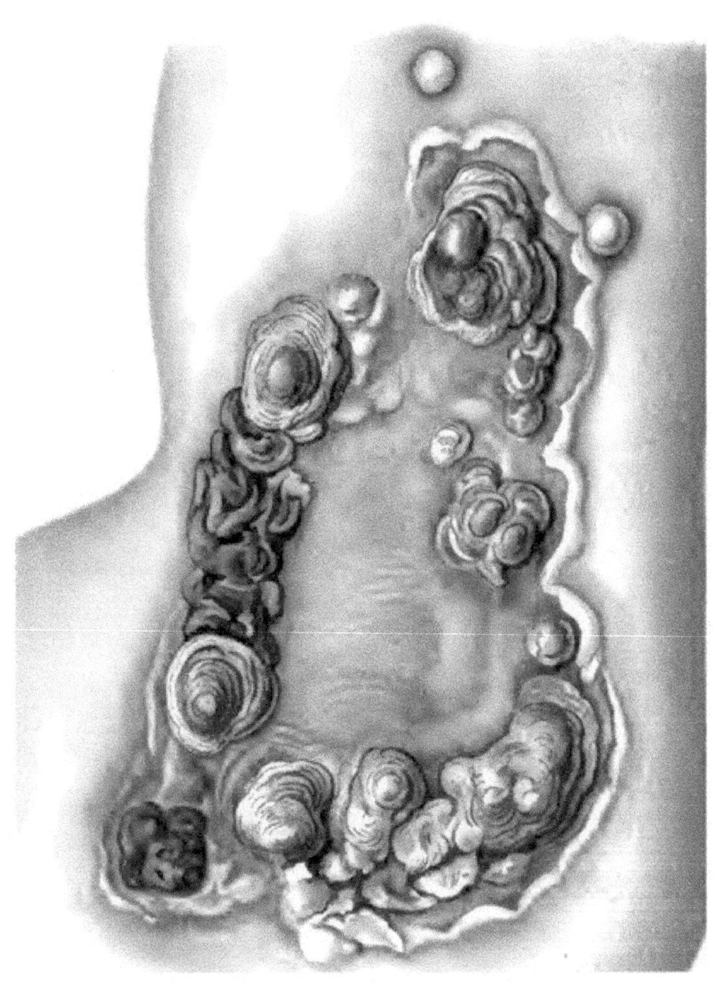

PLATES 50, 50a.

Serpiginous, Gummy Ulcers.

F. P., 28 years old; admitted Feb. 10, 1896. The patient became infected with syphilis in her eighteenth year, and was at that time treated with inunctions. She had no relapses until fifteen months ago, when a tumor began to make its appearance in the lateral cervical region, and was followed by tumors in other parts of the body, all of which softened and broke down. The ulcerations were cured by inunctions, leaving scars.

The ulcers now seen on the right thigh and on the left leg developed six weeks ago.

Present Condition.—There are no alterations in the genitals at the present time. The glands in general are enlarged. On the trunk and extremities are seen numerous scars of varying size, some pigmented, some white, which, from their shape, evidently represent the remains of serpiginous ulcerations. Here and there, especially in the mammary region, along the costal margin, over the head of the right humerus, on the flexor and extensor surfaces of the left upper arm, on both sides of the neck, and over the eyebrows localized eruptions are seen. The papules are livid red, distinctly raised above the level of the skin; the edges are everted and covered with several superjacent, dirty brown scabs, some of which, especially on the left arm, attain the size of a penny (see Plate). The center is occupied by white and brown scars; as they approach the periphery the color changes to a reddish hue, the scars become puckered, and finally merge into a raised zone of infiltration. This infiltrated margin is composed of single nodules, closely crowded together and merging into one another. A few of the nodules are covered with very thin crusts and scales; the older ones, on the other hand, are covered with several layers of crusts. On the lower third of the left leg is a sharply circumscribed ulcer, about as large as a penny, with discolored floor. On the inner margin of the right thigh, above the knee, another similar ulcer.

The mucous membrane of mouth and throat is intact.

Treatment.—Local application of white-precipitate ointment; inunctions of ℥jss ung. ciner. The skin-lesions disappeared rapidly. Patient was discharged before the end of the cure, at her urgent request.

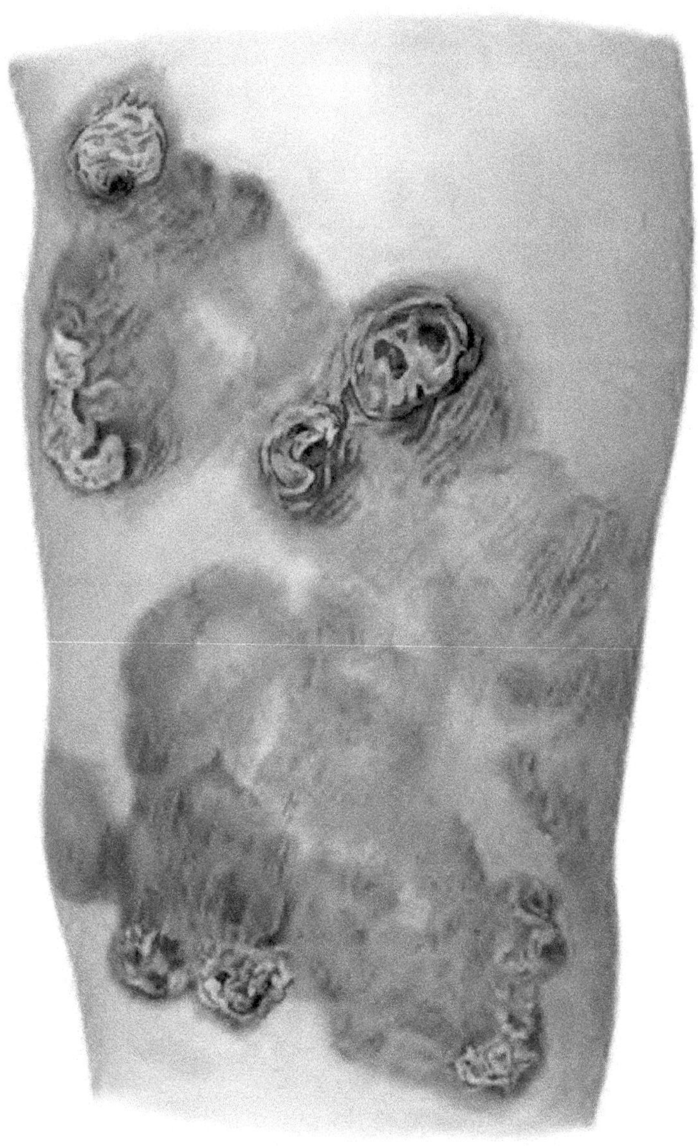

Lith.Anst.v. F Reichhold, München

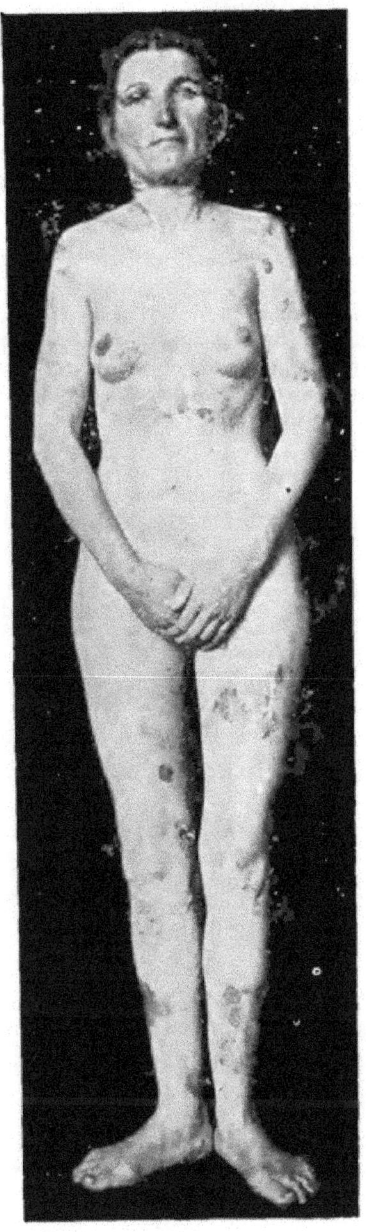

PLATE 51.

Serpiginous, Gummy Ulcers of the Right Calf.

P. T., 30 years old; admitted April 20, 1896. Four years ago the patient contracted a disease from her husband, who was then suffering from an eruption. The disease began in the left tonsil. The cervical glands became swollen; later she was troubled with an eruption and with headache. Since that time the patient was several times treated for various manifestations of the disease, but never continuously. Eleven months ago she began to notice nodes on the calf of the left leg, which soon ulcerated. She has had seven children; the last one was born at term, but she says it is afflicted with an eruption.

Present Condition.—On the calf of the right leg is a group of typical, serpiginous, gummy infiltrations and ulcers, surrounding a central scar, the remains of old ulcers; about the periphery circular and elliptical ulcers of varying size, with fairly well-defined edges and the base covered with granulations and detritus. Similar ulcers are seen above the left knee. In the right groin and on the right labium majus the remains of infiltrations and the scars of papules can still be seen. Some slight pigmentation can be made out on the trunk and extremities. The inguinal glands are only slightly, the cervical glands typically enlarged.

A cure was effected by local applications of red-precipitate ointment, internal administration of potassium iodid, and a course of twenty-one inunctions.

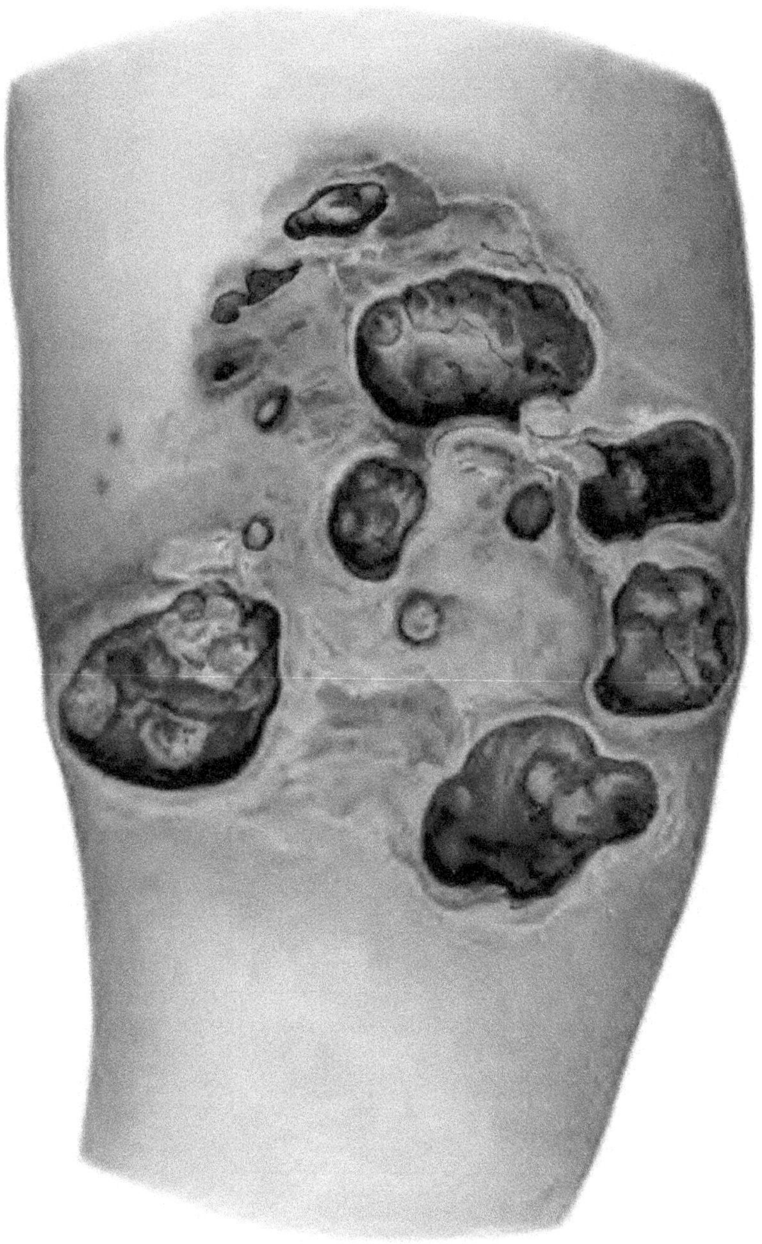

Lith.Anst. v. F. Reichhold, München

PLATE 52.

Cutaneous Gumma on the Dorsum of the Foot. Gumma of the Pharynx.

B. M., 37 years old, married; admitted Dec. 19, 1895. Said to have had a nasal voice ever since her eighth year. Nothing in the history has any special bearing on the origin of the disease. The patient has had six children, all of whom died in infancy from intercurrent diseases.

Present Condition.—The soft palate and uvula are entirely wanting, so that the nasopharyngeal cavity extends high up into the roof of the mouth. The posterior wall of the pharynx presents a yellowish, discolored, ulcerated area about as large as a penny.

On the dorsum of the right foot, corresponding in position to the fourth and fifth metatarsophalangeal articulations, is an ulcer as large as a dollar, filled with proliferating granulations. The edges, where they exist, are sharply defined and overhang the mass of granulations, so that a probe can be inserted 2 to 4 mm. under the undermined edges. The ulceration extends down to the sheaths of the tendons, although the movability of the toes is unimpaired.

The case was treated surgically.

PLATE 52a.

Ulcerative Gummata of the Pericranium.

K. E., 50 years old, pauper. The patient has been treated repeatedly in a ward during the last three years for severe syphilitic manifestations.

On the right parietal bone is a depression as large as a dollar, at the bottom of which the bone is exposed. The soft parts about the periphery of the ulcer-sore are loosened. On the right frontal bone is a similar, smaller swelling about as large as a penny, and a third one is seen on the occiput. There is also a periosteal gumma on the right tibia. The patient is very weak and emaciated, and has edema in the lower extremities.

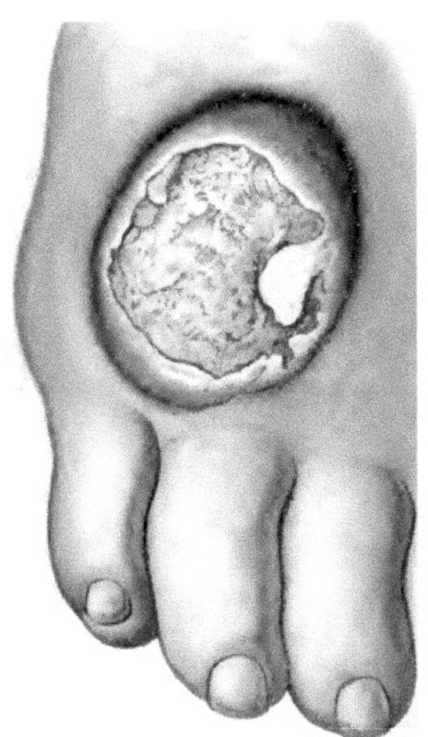

By careful local treatment and general tonics the gummata were gradually absorbed after four months' treatment. The one on the forehead healed in such a manner that the integument united and the bone was covered by granulation- and scar-tissue.

Tab. 52a.

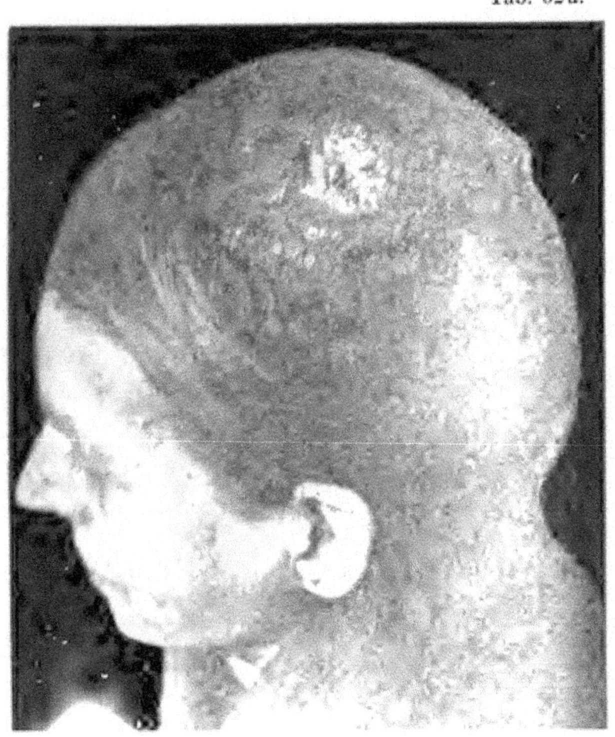

PLATE 53.

Gumma in the Glands of the Neck, with Destruction of the Integument.

B. K., 32 years old, servant-girl; admitted Oct. 5, 1895. There is no history of hereditary disease. Patient says she occasionally suffers from nocturnal headache. Two years ago she was treated in a throat clinic, and a year ago in a surgical clinic for an ulcer on one of the lower extremities. Has never been subjected to a general antisyphilitic treatment and denies any knowledge of the disease.

Present Condition.—The patient is well nourished, somewhat pale. Internal organs normal; has never been pregnant; menstruation regular.

No alterations can be made out on the external labia or in the rest of the genitals. The skin of the neck and throat is the seat of a typical leukoplasia. On the external surface of the left calf a circular, depressed atrophic scar; above the left external malleolus a scar measuring about 2 square cm., adherent to the bone, with irregular, circular, and elliptical margin. The soft palate and uvula partly destroyed and disfigured by scars.

The inguinal glands are hard, and present a multiple swelling; the axillary glands on both sides are even more distinctly enlarged, those above the bend of the elbow only slightly so.

All the glands in the neck, especially those in the left submaxillary and supraclavicular regions, are swollen to the size of pigeons' eggs and hard and resistant to the touch. Two elliptical ulcers about 1 cm. long, corresponding to a submaxillary and a supraclavicular gland, are seen on the left side of the neck; the ulcers have broken through the skin; the edges are steep; the floor of the upper one is covered with necrotic tissue, that of the lower one with yellowish-white pus. Over the other swollen glands the skin is loose and freely movable.

The patient received 24 grains of potassium iodid per diem; the sores were dressed first with iodoform and later with gray plaster. The ulcers healed and the glands were reduced in size, so that the patient was discharged cured after thirty-eight days, the neck having regained its normal outline.

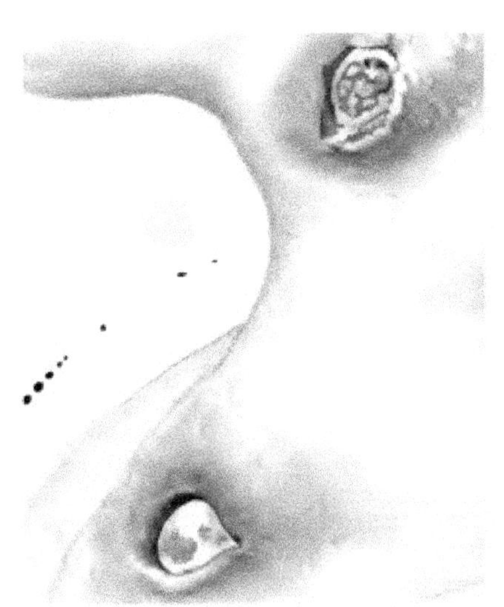

PLATES 54, 54a.

Gummatous Ulcers of the Skin and Inguinal Glands.

W. K., 46 years old, tailor; admitted Jan. 21, 1895; died Jan. 7, 1896. In 1892 the patient had an eruption covering almost the entire body, most abundant on the trunk, for which he used a white ointment. The nature of the disease was not known to him.

Present Condition.—Pigmented and atrophic cutaneous scars can be seen on the entire body, due to the above-mentioned eruption. On the inner aspect of the left thigh are *five* ulcers ranging in size from a penny to a half dollar, penetrating below the skin, the base presenting some granulation and much purulent necrotic tissue.

In the groin is a large, oval wound, corresponding to a broken-down superficial gland. The other lymphatic glands are hard, but little enlarged. The patient is pale and emaciated; he keeps the lower extremity flexed at the hip and at the knee. The joints themselves are unaffected. His psychical condition is normal; his intelligence, however, is of a low order. The gummatous ulcers were not affected by potassium iodid and local treatment with iodoform, so that inunctions were ordered.

Mar. 24. After thirty inunctions the wounds showed active granulation and a border of scar-tissue, but the inunctions had to be discontinued on account of severe gingivitis and abnormal proliferation of the epithelium at the margin of the tongue and in the mucous membrane of the cheek, opposite the alveolar border. In spite of careful nursing neither the ulcers themselves nor the general condition of the pale, torpid patient improved, so that at the end of September the wounds were but little reduced in size.

In the beginning of October a gland in the right inguinal region became swollen. The integument became inflamed and ulcerated, and a thin, bloody secretion was discharged. After three weeks the wound improved somewhat, so that only a part of the degenerated gland and a slight granulation could be seen in the floor (colored plate).

Oct. 21. Erysipelas developed from the right inguinal fold to the middle of the thigh; incipient bed-sore in the right sacral region.

The erysipelatous inflammation subsided upon the application of compresses of aluminum acetate solution and with proper care as to diet and change of position in bed; but on Oct. 24th the patient still complains of severe pain in the right

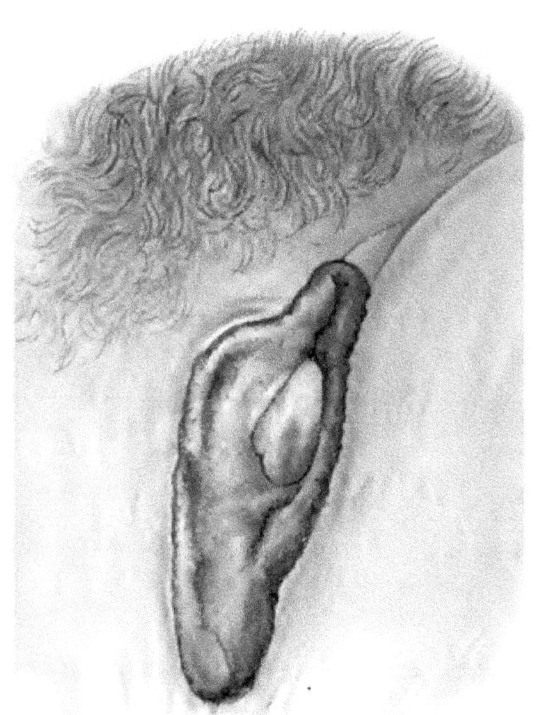

hip, which is found to be red and inflamed. This condition is also relieved in six days with sodium salicylate internally and cold compresses.

Ophthalmoscopic Examination. — Discoloration of the pupil, incipient atrophy.

The patient complains of severe pruritus. Urine contains no sugar, and only small quantities of albumin. The sediment contains many leukocytes and bladder-epithelium, but no renal elements.

Bronchitis affecting the larger bronchi of the entire lung. The fever is moderate; in the morning it falls almost to normal, in the evening it rises to 38.5° C. The patient gradually sinks into a very low state, and has to be roused to take nourishment; he barely understands what is said to him and immediately relapses into a stupid state. He died Jan. 7, 1896.

Autopsy.—Body small, very much emaciated. In the left inguinal region the skin is destroyed over an area as large as the hand.

In the right groin a similar ulceration, about as large as a dollar; on the posterior aspect of the left thigh is a larger scar, freely movable over the muscle; a smaller one on the external surface of the left thigh and over the head of the fibula.

The skull is thin, very prominent in the suboccipital region. The dura mater and soft layers of the meninges present nothing abnormal. The surface of the brain is somewhat flattened. The cortex slightly narrowed. Brain-substance edematous. Nothing abnormal in the vessels at the base.

Both *lungs* emphysematous; atrophied (poor in substance); no adhesions. The bronchi of the lower lobes contain a purulent secretion.

Heart small and contracted; subpericardial layer presents some fatty yellow and red discolorations. Myocardium yellowish-brown. Valves and vessels present no alterations.

Liver rather small, convexity increased. The individual lobules over the entire surface are very prominent, their fatty, yellow color contrasting with the reddish hue of the intervening connective tissue. The same picture is presented in cross-section. On the right lobe is a yellowish, calcified nodule about as large as a pea.

Spleen enlarged and flaccid, stroma increased; a slight waxy luster is seen on section.

Both *kidneys* much enlarged, tough, the capsule easily removed. The surface has a waxy appearance; on section a few fine hemorrhagic points are seen; the cortex is increased in width, pale yellow, in marked contrast with the flesh-colored pyramids, and looks distinctly like bacon. Pelvis dilated and filled with fluid containing dark, turbid flakes; the mucous membrane discolored by numerous hemorrhages.

The *bladder* dilated to its utmost, containing clear urine. Mucous membrane strongly injected in places.

On the right side of the neck of the penis an atrophic scar.

The mucous membrane of the intestine about the anus is puckered and thickened, and protrudes from the anus.

Over the sacrum a large, irregular ulcer extending chiefly toward the right side.

Diagnosis.—Inveterate syphilis. Cirrhosis of the liver in process of regeneration. Amyloid disease of kidneys and spleen. Atrophy of the heart. Slight atrophy of the brain. General anemia and marasmus.

Tab. 54 a.

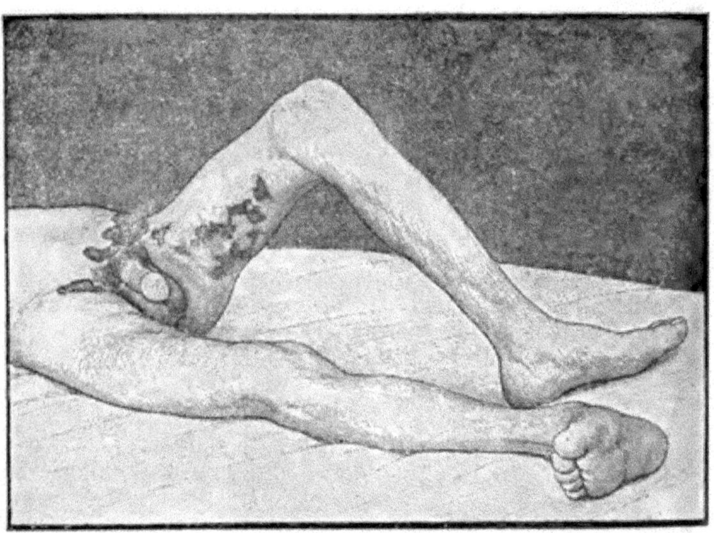

PLATE 55.

Gummatous Disease and Necrosis of the Soft Parts.

H. A., 29 years old, servant-girl; admitted May 9, 1896. The patient had diphtheria and small-pox when a child. In her eighth year she suffered with a disease of the left fibula. The disease in the upper arm began a year ago. Her menses did not begin until she was eighteen years old; has always menstruated regularly. The patient gave birth to a child last August; the child was weakly and died in four weeks of a birthmark, so the patient says.

Present Condition.—The patient is of slender build and emaciated. Teeth are bad. Nose slightly saddle-shaped. The throat marked with scars. The remains of the soft palate are drawn against the posterior wall of the pharynx. Uvula is wanting.

There is a radiating, movable scar over the acromion and clavicle on the left side. It represents the remains of an ulceration which followed the patient's confinement last year. Shortly afterward an ulceration developed on the left upper arm and lasted two months. The contracting, radiating scars which remain are directly adherent to the bone. At the same period a stiffness began to show itself in the elbow-joint. At the beginning of the present year, when the disease on the back of the arm had hardly begun to heal, the ulcer on the front of the arm developed, and soon after that the one on the upper third of the forearm and in the bend of the elbow. The left arm is held in extension, the hand in extreme pronation. Flexure at the elbow and supination much impaired. The upper two-thirds of the radius seem to be thickened, as is also the lower end of the upper arm. The external surface of the upper arm is occupied by an ulcer 9 cm. long and 3 cm. wide, about the periphery of which a little scar-tissue is beginning to form, with here and there a few granulations. The floor consists of necrotic muscle-fibers lying lengthwise, of a dirty yellow color and surrounded by irregular depressions which discharge a scanty secretion. Further down toward the elbow there is a raised wheal, and near its outer margin an oval ulcer about 2 cm. in diameter. A bridge of scar-tissue about 1½ cm. wide

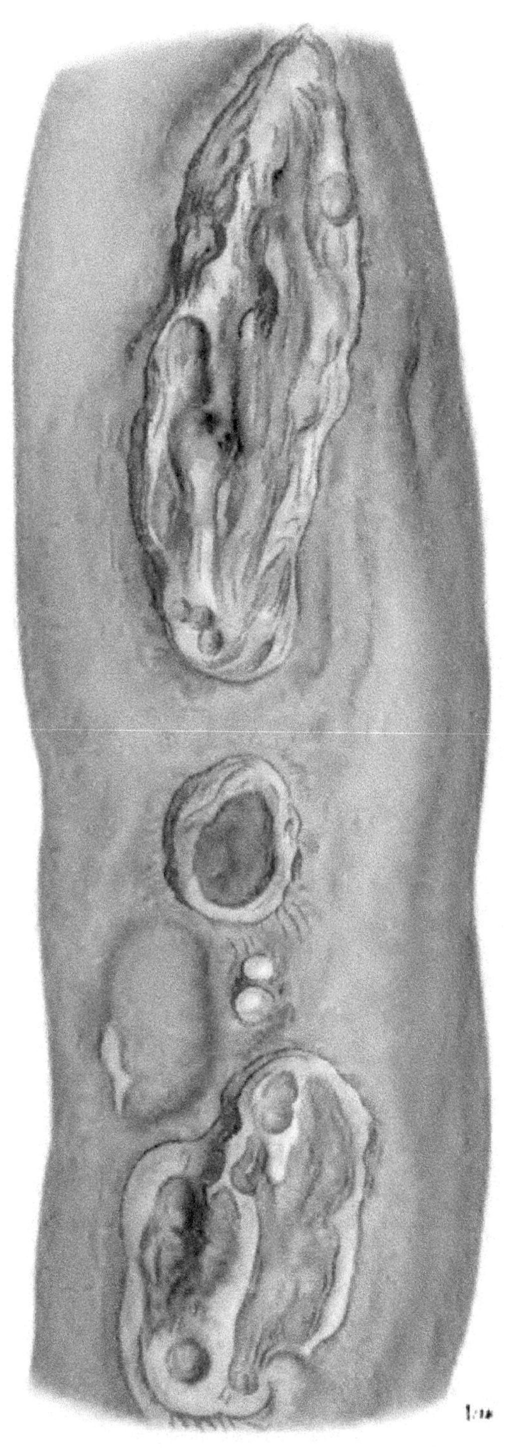

separates this ulcer from the one above it. The floor is formed by the brownish necrotic skin. The subcutaneous cellular tissue, macerated with serum, is exposed for about 3 to 4 cm. from the edge of the dry crust of skin. Below this latter ulcer, close to the wheal, there is a third ulcer with seropurulent floor; divided into two halves by a bridge of skin. Finally, on the upper third of the forearm and in series with this last-mentioned double ulcer, is one about 5 cm. long, the inner half of which, lying toward the radius, is covered with granulations and attached to the radius by scar-tissue. In the outer half of the ulcer is a fragment of skin on the point of desquamation, and brownish subcutaneous tissue, resting on a foundation containing a scant serous exudation, so that the necrotic parts are easily movable over the underlying tissue. This ulcer lies between the radius and the ulna, over the pronator radii teres muscle, and seems to have sprung originally from the radius. In the bend of the elbow is the above-mentioned wheal with the two ulcers near its outer border. The lower third of the radius is enlarged from periostitis; the upper part also thickened in irregular lines in the long axis of the bone. The ulna does not appear to be involved to the same degree. On the other hand, the entire lower third of the upper arm, almost as far as the middle, is enlarged and, with the exception of the large ulcer described, covered with scar-tissue. In some places the bone is reduced in thickness from atrophy; in others it is enlarged from periostitis.

On the outside of the left calf is a scar 10 cm. long, adherent to the bone, the result of the above-mentioned disease of the fibula. The genitals present no alterations. Several white, atrophic scars about the anus probably date from the attack of small-pox. The inguinal, as well as the other glands of the body are reduced in size.

PLATE 56a.

Destruction of the Soft Palate by Gummatous Ulceration.

S. M., 25 years old. The patient does not know how long her disease has lasted. She only began to feel pain in the throat three weeks ago.

Present Condition.—At the edge of the labia majora several whitish, hairless, areolar scars. Similar, reticulated scars about the anus and in the vestibule. In addition, other scars on the inner surface of both labia minora. Inguinal glands hard and spindle-shaped. Almost the entire soft palate wanting, the edges of the wound being covered with shreds of necrotic tissue. The ulceration has invaded the edges of the arch of the palate, so that the upper and lateral boundaries of the isthmus are also involved in the degenerative process.

Cured by the application of fifteen inunctions and 64 g. (℥ij) of potassium iodid internally.

PLATE 56b.

Gumma (on Posterior Wall of Pharynx).

R. R., 39 years old, no occupation; admitted Jan. 6, 1896. History very meager. Patient has been married eleven years; says she has never been pregnant. For the last three years she has been troubled with a "nasal affection;" she says the secretion has a very offensive odor. Regurgitation of soft and liquid food began quite suddenly a week ago. The patient's husband admits that he was infected with syphilis three years ago.

Present Condition.—No remains of a syphilitic infection can be demonstrated on the genitals. At the anus a red scar about as large as a pea. Multiple swelling of the inguinal glands.

The soft palate is destroyed and replaced by a scar; the uvula is entirely wanting. Strands of scar-tissue are attached to the posterior wall of the pharynx, on which there is an elliptical ulcer about 1½ cm. by ½ cm., its long axis corresponding with that of the pharynx. The center is depressed and partly covered with a dry, black scab; near the edges it is slightly flattened. To the right of this ulcer the orifice of the Eustachian tube can be seen. Rhinoscopic and laryngoscopic examination reveals no other alterations.

Voice is nasal. The breath is very fetid.

Treatment.—Inunctions and potassium iodid.

After twenty-five inunctions the ulcer on the posterior wall of the pharynx healed over completely and was replaced by scar-tissue.

Discharged Feb. 11, 1897.

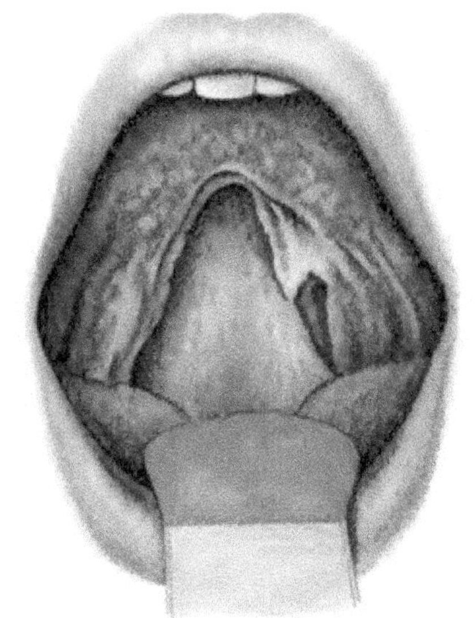

a

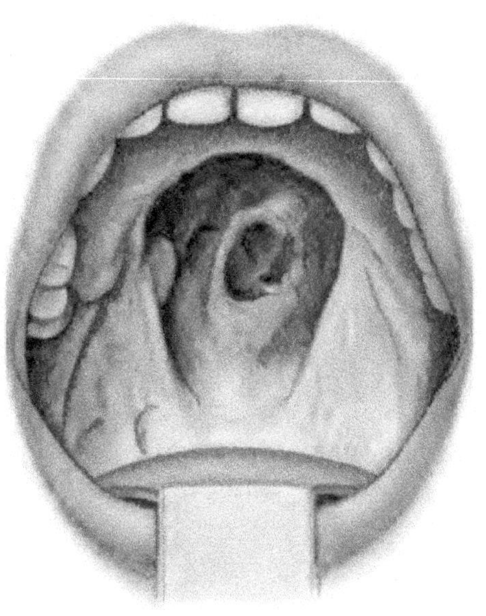

b

PLATE 57.

Gummatous Glossitis.

A. F., 25 years old, servant-girl; admitted Oct. 28, 1890. Under treatment one month. Second attack. Has had an ulcer on the tongue for three months. Five years ago the patient was under treatment three months in the syphilitic ward of the Wiedener hospital for a specific ulcer.

Present Condition.—Ulcer on the right labium majus. No other signs of a former or still existing syphilitic attack either on the skin or in the glandular system. When the mouth is opened wide and the tongue well protruded a swelling is seen covering the entire posterior half of the left side. The tumor is raised 3 to 4 mm. above the surrounding surface; it is hard to the touch and extends through the entire thickness of the organ from its base to about its middle. The surface of the tumor is traversed by an ulcer 3 cm. long. From the middle of the tongue, extending almost to the tip, there is another tumor consisting of a number of nodular infiltrations, showing necrotic decay in three places.

Treatment. — Potassium iodid. Antiseptic mouth-wash. On Nov. 17th an inunction treatment was inaugurated. The infiltration was absorbed and the ulcers healed. After fifteen inunctions the patient was discharged cured.

PLATE 58.

Papulopustular Exanthema. Hereditary Syphilis.

(Obtained through the kindness of Dr. Braun, of the Foundling Asylum.)

S. K., about 4 weeks old, weight 5¾ pounds; admitted July 8, 1897. Marasmus marked. Suffering from bronchitis and intestinal catarrh. The skin is pale and wrinkled, and thickly covered with a syphilitic eruption. The forehead and mouth, and also the trunk and extremities, are the seat of papules with pale-red border, or vesicles containing a small quantity of serous exudate, with flaccid, partly degenerated epidermis. Died after twenty-four hours.

Autopsy.—General tabes, bronchitis, lobular pneumonia on both sides, enlarged spleen, hepatitis, gastro-intestinal catarrh, syphilitic osteochondritis at the epiphyses of the tibiæ.

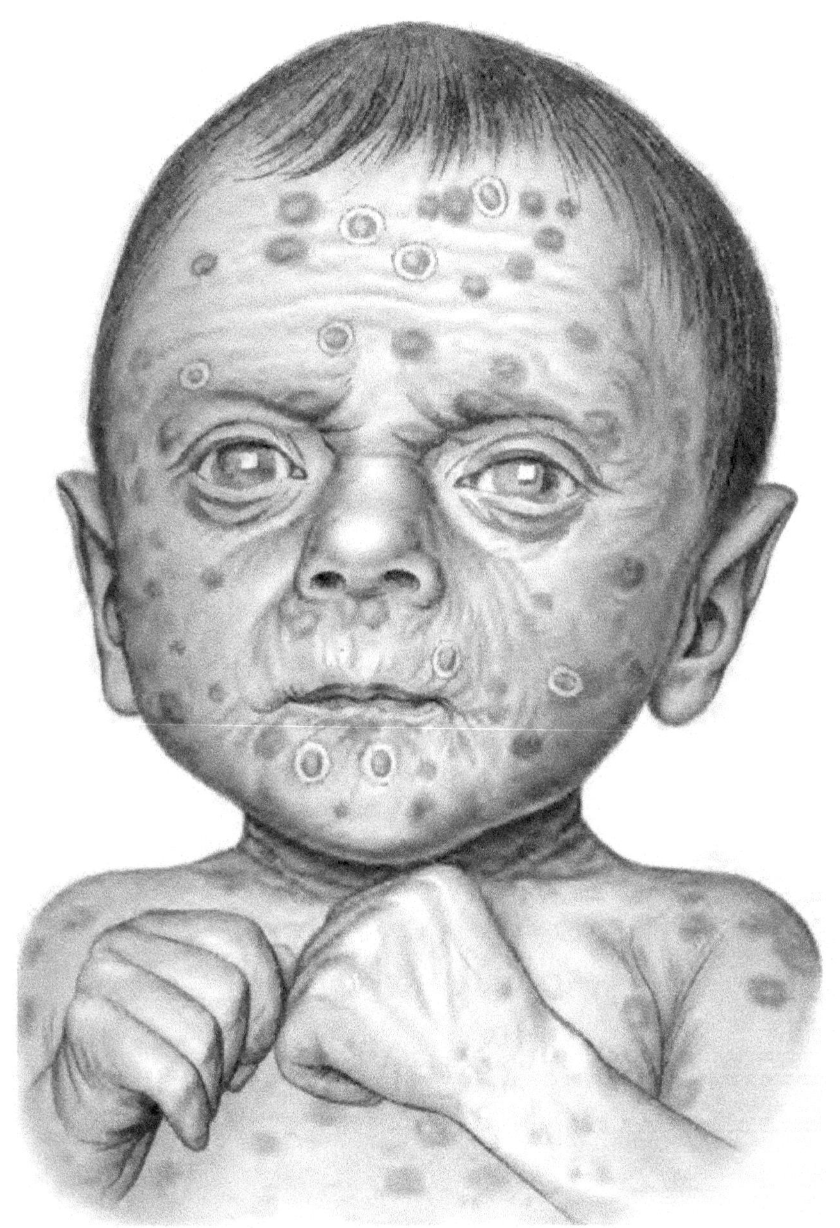

PLATE 59.

Papulovesicopustular Exanthema. Hereditary Syphilis.

On the legs and on the soles of the feet papular and vesicular infiltration and vesicles, ranging in size from a lentil to a pea, containing pus and surrounded by an inflammatory halo.

F. J., born June 9, 1897; admitted to the Foundling Asylum June 10, 1897. Weight at the time of admission 8½ pounds. Mother apparently healthy.

On June 15th an eruption, consisting principally of papules, appeared on the palms of the hands and soles of the feet. On the following day the extensor surface of the lower extremities, the nates, and the back were also covered. Some vesicles and pustules are seen among the papules. The nose is not affected.

Later on symptoms of bilateral lobular pneumonia and gastro-intestinal catarrh appeared. The baby's weight gradually fell to six pounds. Died June 26, 1897.

Autopsy.—Lobular pneumonia in the lower lobes of both lungs, infiltration of the liver, enlarged spleen, gastro-intestinal catarrh, no osteochondritis.

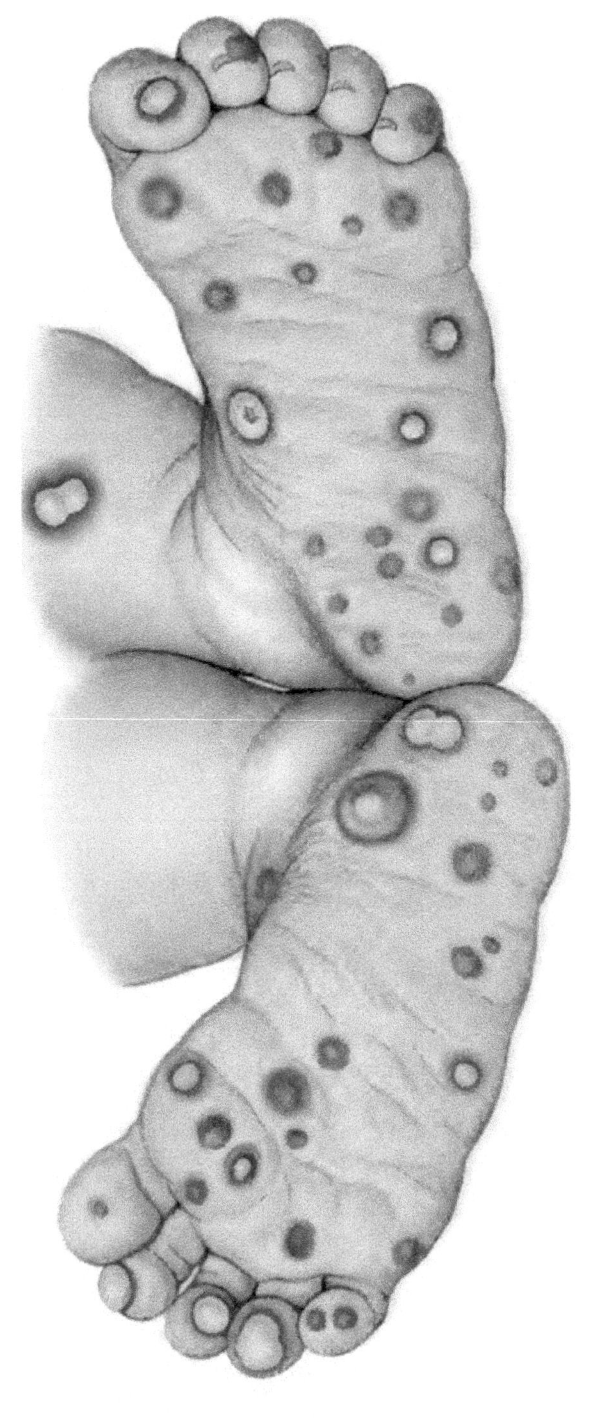

PLATE 60a-c.

Hereditary Syphilis.

(Parenchymatous keratitis; syphilis of the bones of the nose; Hutchinson's teeth.)

K. A., 20 years old, servant; admitted June 5, 1897, to Professor Bergmeister's ward for diseases of the eye. The patient says his right eye was diseased nine years ago. The disease in the left eye began two weeks ago. Photophobia and excessive lachrymal secretion.

Present Condition.—The entire body, though of normal build, is distinctly puerile for its age of twenty years; the bones, especially those of the extremities, are soft and greatly enlarged at the joints; the genitals are infantile and the pubic hair very scantily developed. The compact, dolichocephalous skull presents a marked contrast to the soft bones of the extremities. The bridge of the nose is sunken and saddle-shaped (see *Black Plate* 60c). The lips, especially the upper one, heavy and hypertrophied. Upon inspection of the nasal cavity the cartilaginous as well as the bony portion of the vomer is found to be eroded by the destructive ulceration; the probe strikes upon a rough, bony surface. When the upper lip is raised, it is seen that the necrosis has invaded the upper maxillary bone, a narrow bridge of alveolar process being necrotic as far as the margin and limited in front by a remnant of gum and by granulations. The teeth present the type of congenital syphilis described by Hutchinson: the irregular arrangement and chisel-shape, with notches in the margin.

Ophthalmoscopic Examination (by Professor Bergmeister). —*The right eye* diverges; it is at present free from irritation; in the center it shows traces of parenchymatous cloudiness. The pupil retracts promptly.

The left eye: ciliary irritation; the upper margin of the limbus is swollen, the adjacent zone of the cornea dim and granular. A dense parenchymatous cloudiness encroaches upon the cornea from under the upper margin of the cornea.

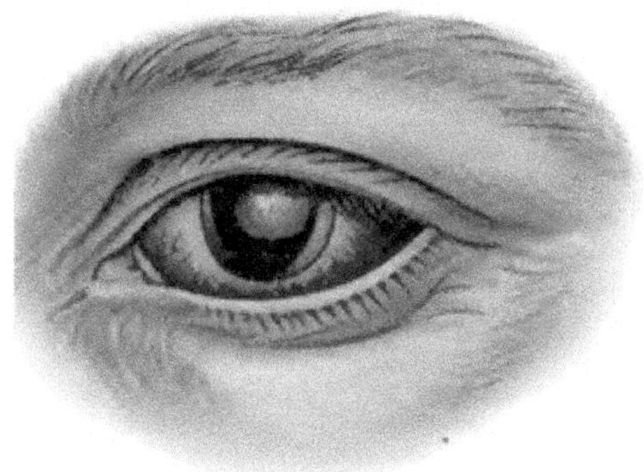

a

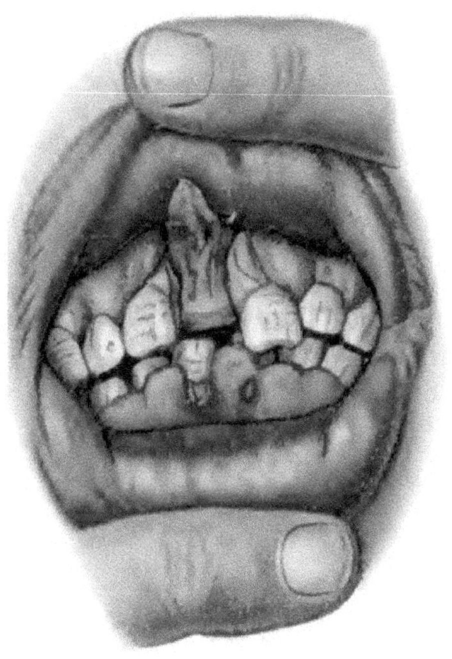

b

Treatment.—Atropin; inunctions; potassium iodid.

June 11. Congestion in the left eye has disappeared; slight turbidity in the upper half of the cornea. Discharged cured.

Tab. 60 c.

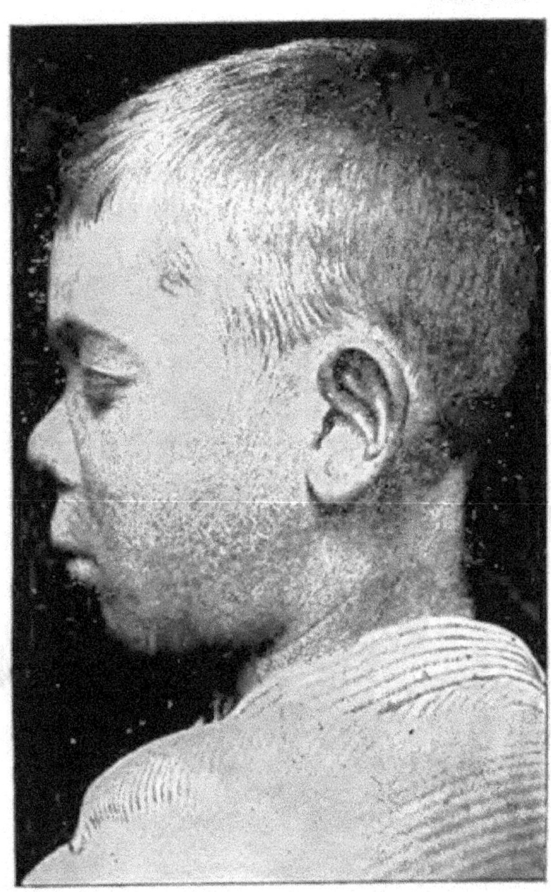

PLATE 61.

Venereal Ulcers in the Foreskin and on the Head of the Penis.

V. E., 26 years old, machinist; admitted Sept. 17, 1896. This is his first attack. Patient had his last coitus two weeks ago; he first noticed the ulcers ten days ago.

On the left side of the inner layer of the prepuce are two large venereal ulcers, surrounded by a zone of moderate inflammation. The edges are broken down and the floor discharges freely. Five smaller ulcers, about as large as lentils, below the larger ones, and one ulcer on the head of the penis represent the second generation of these auto-infective sores.

The ulcers healed upon application of formalin-gelatin and cauterization with carbolic acid, leaving scars.

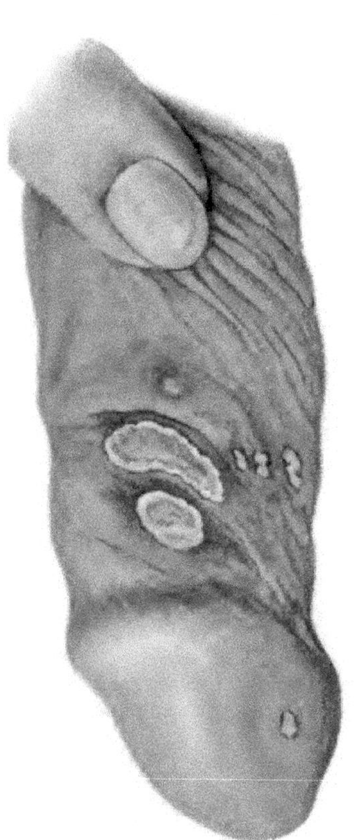

PLATE 62.

Contagious, Coalescent Venereal Ulcers in the Skin of the Penis. Adenitis of the Right Inguinal Glands.

K. J., 29 years old, coachman; admitted Nov. 26, 1896. Patient says it is his first venereal attack; first noticed the sores two weeks ago; last coitus three weeks ago.

Present Condition.—In the skin of the penis, about its middle, is a large venereal ulcer formed by the confluence of several smaller ones; a second smaller one nearer the end of the penis, separated from the first by a slender bridge of skin. Both ulcers have penetrated beyond the skin; the edges are sloping and irregular in outline; the secretion of pus is copious, and the ulcers show a tendency to spread by undermining the adjacent skin. The surrounding tissues are inflamed, the lymphatics of the dorsum being red and distinctly visible. In the right inguinal region there is a tumor as large as a child's fist, red in the center, painful, and slightly fluctuating.

Treatment.—Airol powder locally. Operation for the adenitis. Cured.

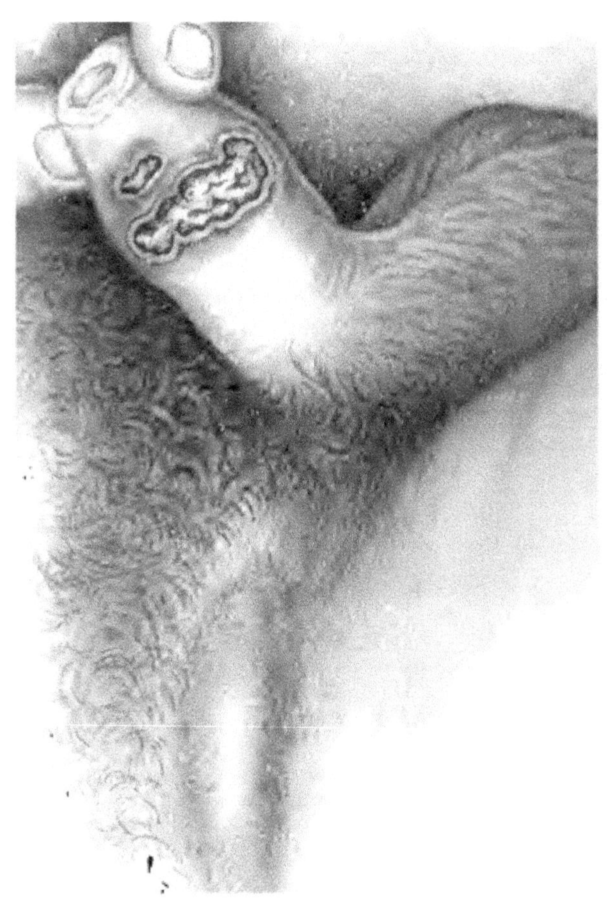

PLATE 63.

Paraphimosis from Venereal Ulcer on the Foreskin. Inflammatory Edema. Suppurative Adenitis in Both Groins.

S. F., 21 years old, glove-maker; admitted Jan. 30, 1897. Last coitus seven weeks ago. Patient first noticed the ulcers four weeks ago; the adenitis, two weeks ago.

Present Condition.—Prepuce swollen and inflamed, retracted; reposition impossible. Large, suppurating ulcers in the region of the frenum; a smaller one on one of the folds of the retracted edematous prepuce. Inguinal glands swollen on both sides; the skin over them inflamed, and beginning to ulcerate on the left side. Vesicular eczema of the pubic region from the abuse of gray ointment.

Treatment.—Sulphate of copper baths, iodoform powder (for the ulcers). Operative removal of the suppurating inguinal glands.

Later on, the ulcer on the frenum proved to be indurated, a papular syphilide appeared on the trunk, and the patient had to be subjected to inunction-treatment.

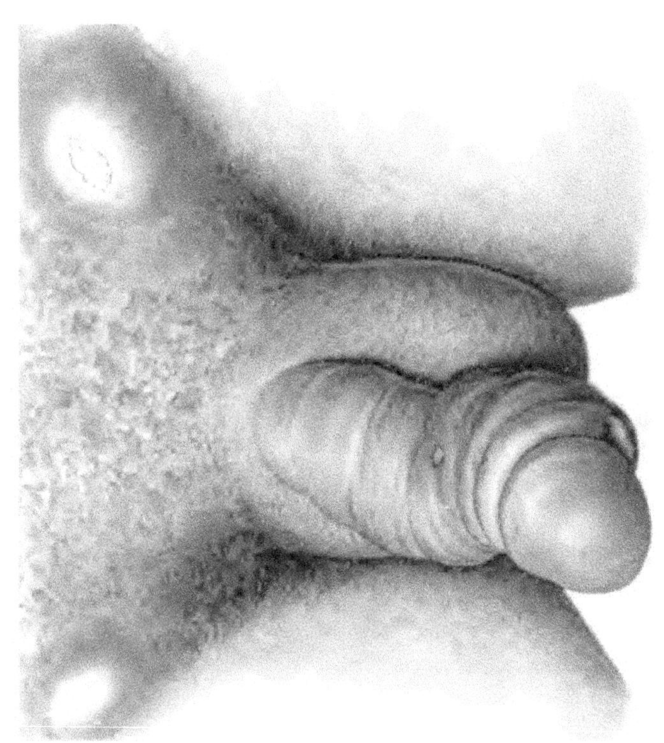

PLATE 64.

Suppurative Lymphangitis of the Dorsum Penis (Bubonulus Nisbethi), with Necrosis of the Integument.

N. B., 22 years old, locksmith; admitted Feb. 1, 1897. The ulcers appeared three weeks ago. Contraction of the foreskin and the swelling on the dorsum penis began eight days ago. Last coitus two months ago.

Present Condition.—Preputial sac swollen and very much inflamed. A purulent secretion flows from the constricted opening. About the middle of the dorsum penis is a hemispherical tumor projecting above the surface, about as large as a walnut. The exposed surface is about as large as a penny, and presents a dark-brown discoloration. The tumor fluctuates. In front the necrotic scab is beginning to separate from the surrounding inflamed tissue. On pressure thin pus oozes from under the crust. Multiple swelling of the inguinal glands, especially of the right side. After a few days the abscess was evacuated, the necrotic covering fell off, and a large ulcer was exposed, the upper margin of which, lying toward the pubic region, appeared undermined, so as to simulate a fistula, while the floor was covered with a copious purulent secretion.

Cured with sulphate of copper baths and iodoform powder. Paraphimosis removed by operation.

PLATE 65.

Abscess of the Left Gland of Bartholin.

L. J., 19 years old, prostitute; admitted March 10, 1896. Patient had no knowledge of her gonorrhea. Pain and swelling began six days ago.

The left labium majus is converted into a painful inflammatory tumor as large as a child's fist, red throughout its whole extent; the skin toward the internal border thin and of a livid hue, with a distinct fluctuation beneath it. The tumor has pushed aside the right labium majus and the left genitocrural fold.

Treatment.—Incision of the abscess.

PLATE 66.

Gonorrhœa Cavernitis.

J. R., 22 years old, confectioner. Under treatment from Feb. 12 to April 20, 1897. Had been ill eight days before admittance to the hospital; last coitus two weeks ago. Acute gonorrhea.

In the course of the treatment in the hospital a marked swelling developed on the under surface of the penis, which was very painful on pressure.

The member appears bent, the concavity looking upward. When fluctuation appeared to the right of the raphé of the penis, an incision was made and a moderate amount of creamy pus was discharged.

Drainage. The wound healed nicely. Patient was discharged cured.

Tab. 66.

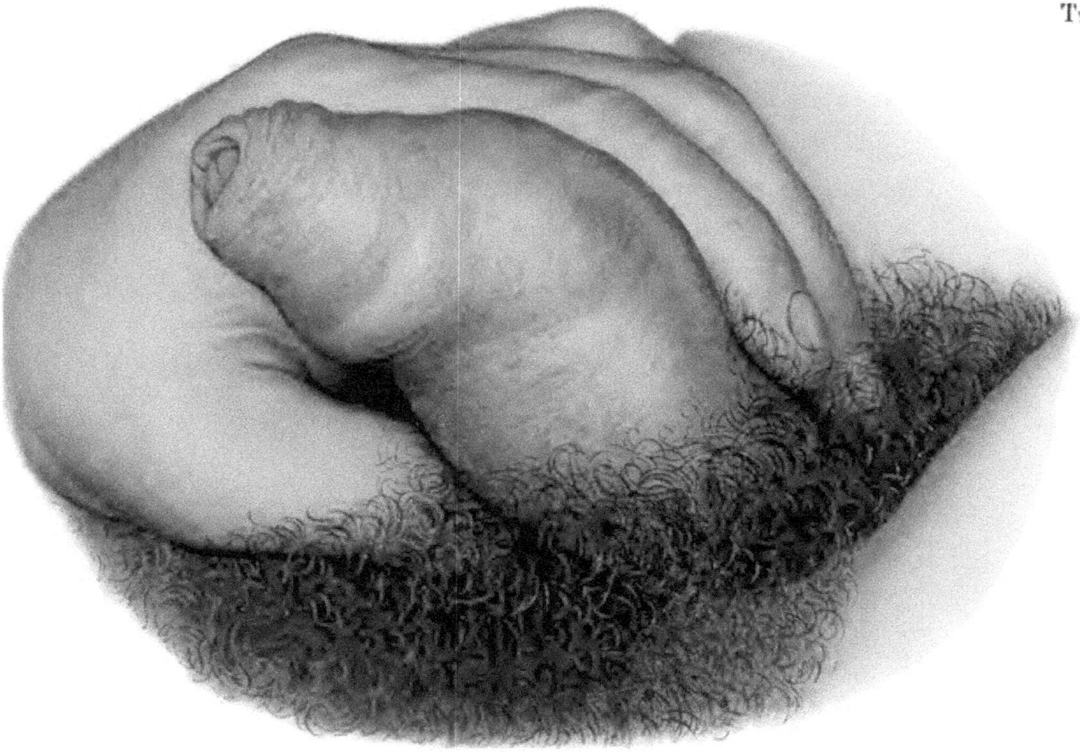

PLATE 67.

Condylomata Acuminata.

G. S., 24 years old, servant-girl; admitted Aug. 18, 1896. The patient says she has had a discharge for five months. The proliferating growth on the genitals began to develop two months ago.

Present Condition.—The labia majora, the perineum as far as the anus, and the region extending to the genitocrural folds are covered with a massive tumor composed of wart-like, nodular papillomatous proliferations. The surface is macerated in places, in others covered with a layer of grayish-white hypertrophied epidermis, and presents here and there isolated areas of bright-red discoloration. When the labia majora are held apart, the labia minora and vestibule appear much inflamed and covered with isolated and coalescent papillomatous proliferations. Urethral gonorrhea. Purulent discharge from the cervix of the uterus.

Treatment.—Removal of condylomata under chloroform anesthesia.

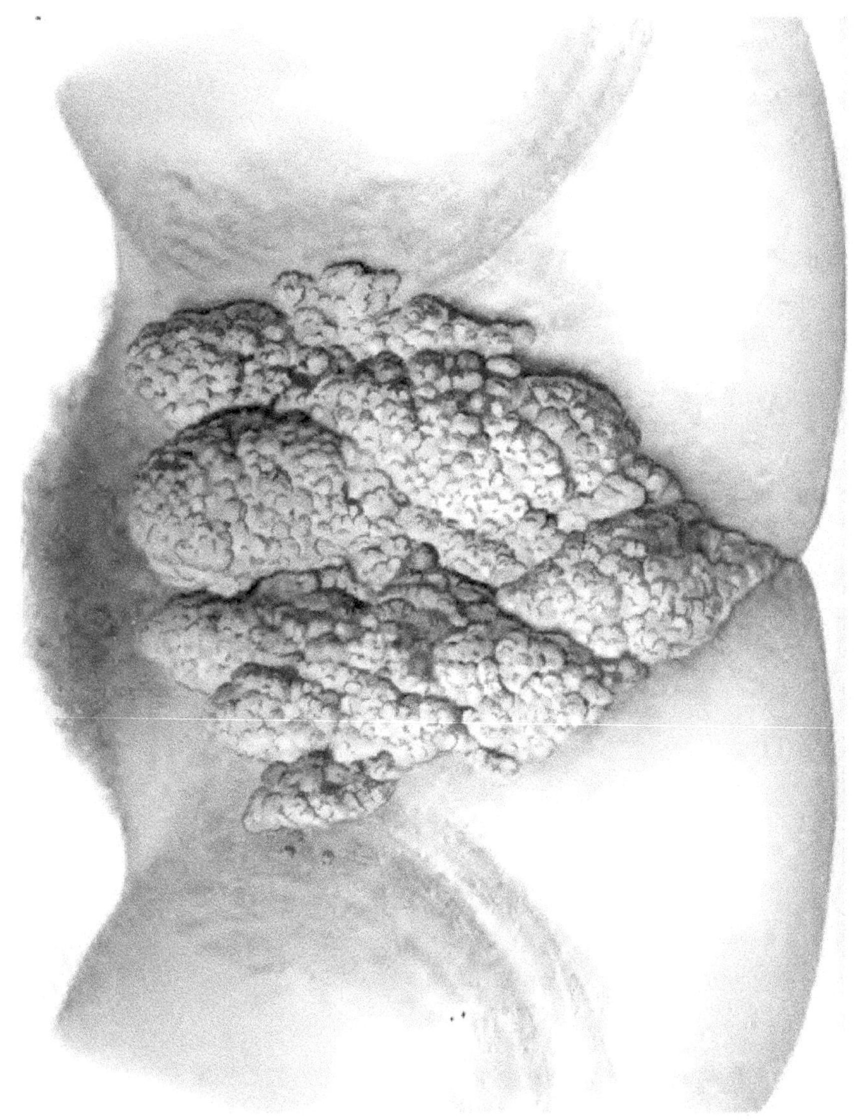

PLATE 68.

Condylomata Acuminata on the Coronary Sulcus and on the Inner Layer of the Foreskin, which is Inflamed and Necrotic along the Left Border.

(The illustration is a copy of the original by Elfinger in the collection of the hospital.)

The left border of the foreskin has become necrotic from pressure and fallen off; the necrosis has also invaded the underlying tissue, which is now exposed. The right portion of the prepuce is displaced to the right and turned back. The space within this expanded preputial sac is occupied by the glans and the surrounding mass of condylomatous proliferations. The latter are covered here and there with a greenish, discolored pus.

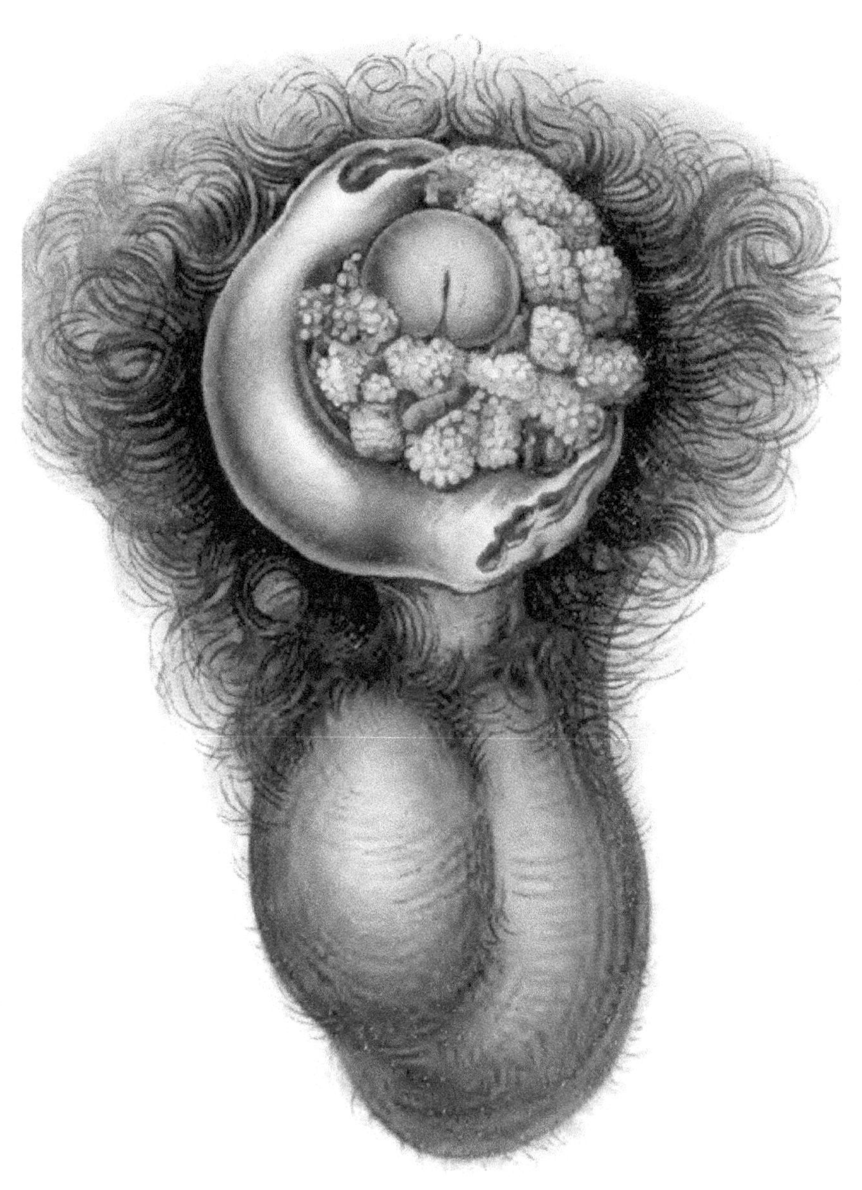

PLATE 69.

Condylomata Acuminata at the Os Uteri.

Cz. A., 19 years old, prostitute; admitted Oct. 12, 1896. A month ago the patient was discharged from a hospital, where she had been treated for gonorrhea and condylomata acuminata. She says the condylomata returned four days (?) ago.

Present Condition.—Acute urethral gonorrhea. Condylomata on the fimbria. Os uteri turned back and flattened; on both lips, especially close to the anterior lip, confluent condylomata acuminata; purulent discharge from the os uteri.

Treatment.—Removal of the condylomata with the curet, after drawing forward the uterus.

Discharged Dec. 22, 1896, cured.

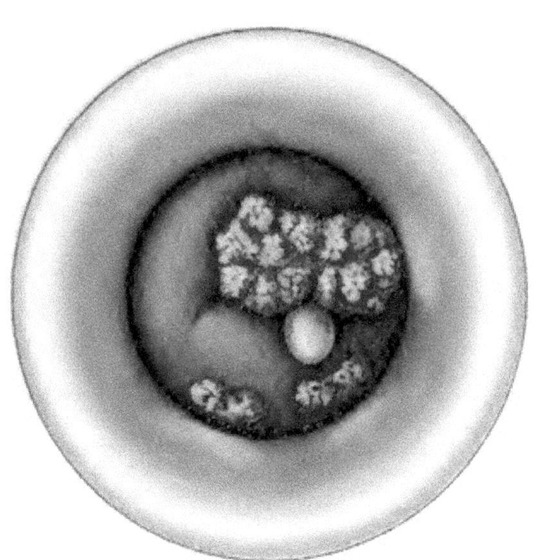

PLATE 70.

Subcutaneous Hemorrhage into the Skin of the Penis.

J. R., 39 years old, factory-hand. Under treatment from May 9 to 19, 1892.

The patient noticed the present condition of the penis on May 8th, immediately after sexual intercourse. The woman with whom he then performed the act of copulation for the first time was, according to his statement, not a virgin, but was very closely built and he had to make a great effort to effect an entrance. The patient also says that he often has spontaneous attacks of epistaxis, and bleeds very freely after having a tooth pulled or from the least cut (hemophilia).

Present Condition.—Vigorous, apparently healthy man. Nothing abnormal detected in the internal organs. Genitals well developed. Penis slightly edematous, especially the lower portion of the prepuce. The skin about the root is tense and of a dark, purplish hue. Color of the glans normal. The lower portion of the prepuce is converted into a purple, edematous tumor, somewhat larger than a walnut, and pendulous.

Upon local application of cold compresses the extravasation was absorbed, the skin passing through the usual color-changes.

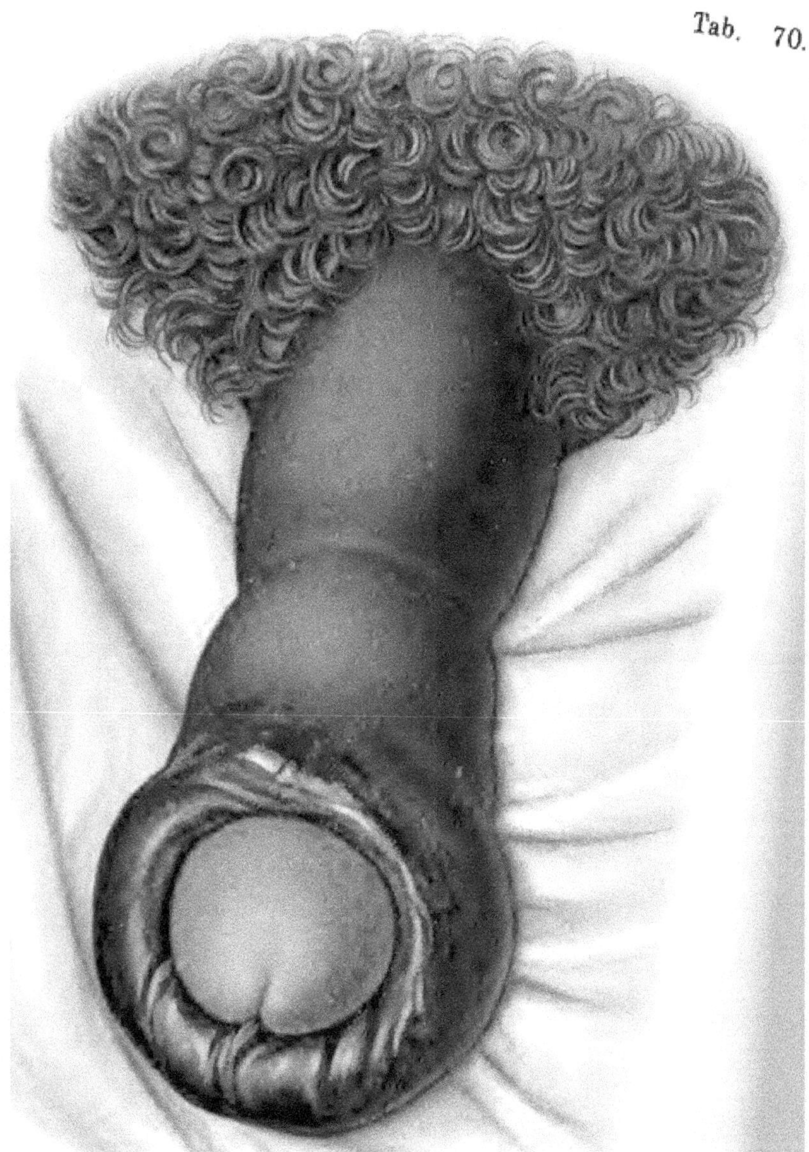

PLATE 71.

Molluscum Contagiosum (Moniliforme [bead-like]).

M. W., 22 years old, washwoman; admitted Feb. 2, 1897. Has had a discharge for two weeks. Gonorrhea.

A large number of pale-red nodules, ranging in size from the head of a pin to a pea, depressed at the center and pierced here and there by hairs, are seen disposed in rows on the outside of both labia majora and extending in a straight line to the buttocks.

Treatment.—Removal of the vegetations with the curet. Discharged Feb. 23, cured.

Tab. 71.

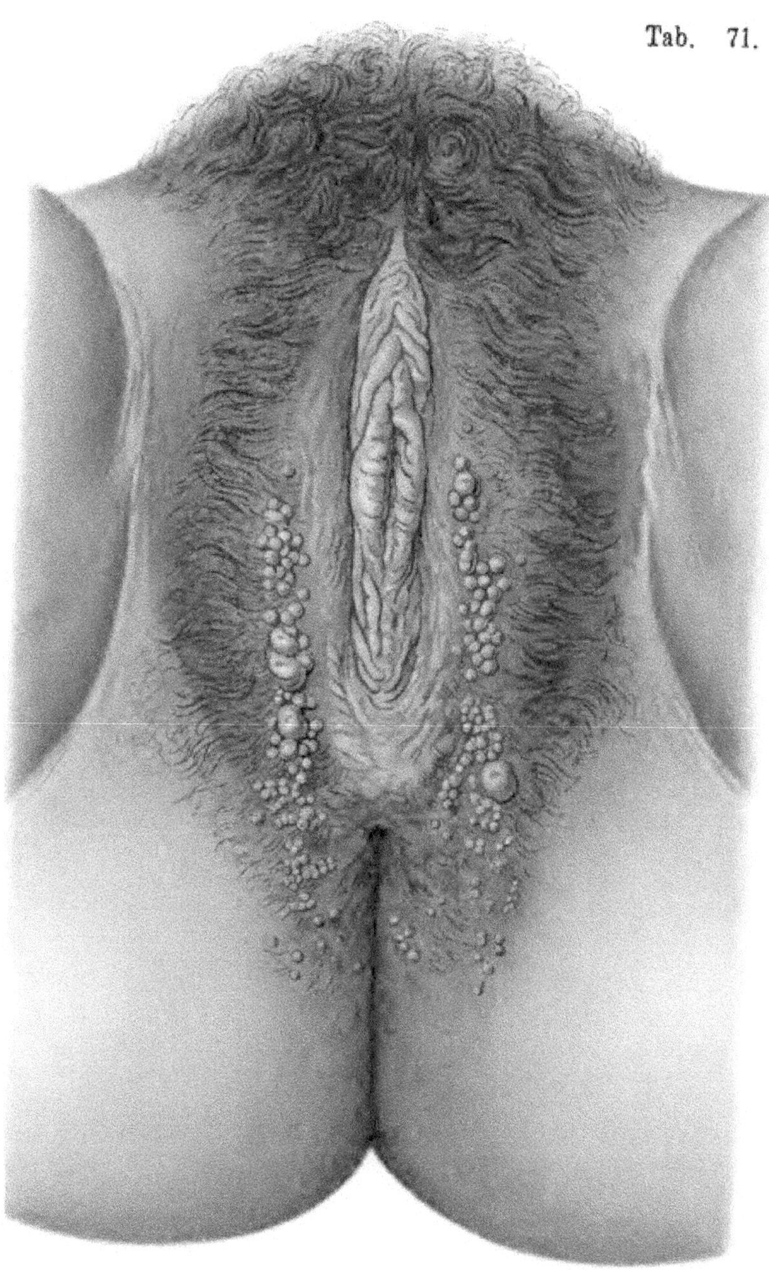

Lith. Anst. v. F. Reichhold, München

INTRODUCTION.

SYPHILIS, also called *lues venerea*, is a constitutional disease, and is classed among the *chronic infectious diseases.*

It is probable that syphilis, like tuberculosis, glanders, etc., is caused by a micro-organism, but the *specific bacillus* has not as yet been isolated.

We cannot, in this short abstract of the pathology, enter into a discussion of the various attempts that have been made to explain the nature of the syphilitic virus, nor shall we even mention the theoretical descriptions and interpretations of the pathological products, which are based on the mere assumption of the existence of a micro-organism and its toxins.

Syphilis can be transmitted from an infected to a healthy individual free from syphilis—*acquired syphilis.* It is, therefore, a contagious disease which follows inoculation with the virus and infects the entire organism, giving rise to various symptoms.

In addition to this mode of infection, syphilis has the property of being transmitted from parent to offspring—*hereditary syphilis.*

Depending upon these two modes of infection, the disease presents in its subsequent course marked variations, so that the two varieties have for a long time been treated separately. We shall therefore adhere to the customary method in our short presentation of the pathology, and shall begin with acquired syphilis. Hereditary syphilis will be discussed later in a separate section.

It shall be our task to study the morbid symptoms produced by syphilis, and to endeavor in this way to acquaint ourselves with the nature of the disease.

Acquired syphilis is transferred to a healthy person sometimes mediately, but in the great majority of cases immediately. A focus is formed at the point of inoculation, and from this focus the virus penetrates into the organism. In order that this may take place, it is necessary, first, that the virus multiply at the point of entry; and, secondly, that it be diffused through the body by means of the lymph- and blood-channels.

The latter process is gradual, and cannot be demonstrated by any known method of examination; we are therefore limited in our observations to the local manifestations at the point of entry of the poison, and to the symptoms immediately following, until other morbid symptoms make their appearance in other parts of the body. It is well to bear in mind, however, that the process of multiplication and extension of the virus through the organism is probably a continuous and uninterrupted one. The patient soon begins to complain of more or less subjective disturbance, and visible as well as palpable objective changes in the integument make their appearance, so that a complete general infection can be definitely recognized a few weeks after infection.

For the sake of clearness and convenience, rather than in accordance with the actual conditions, the syphilitic phenomena are divided into (1) a *primary stage*, comprising the initial local symptoms produced by the infection; (2) a *secondary stage*, beginning with the first appearance of the general symptoms; and (3) a *tertiary stage*, in which nodular formations, so-called gummatous neoplasms, develop. Some specialists base this classification on the clinical appearance of the symptoms; others, on the period at which they are observed to occur. As a matter of fact, the division does not correspond with the actual nature and course of the disease; for, although there are so-called latent or intermission-periods during which the organism is apparently free from the symptoms of the following group, we know that the virus continues to live and multiply within the body, and sooner or later gives rise to renewed symptoms. Perhaps it would be more exact to designate the primary and secondary symptoms together as the *irritative stage*, as Virchow did long ago, separating them from the later gummatous neoplasms and degenerations of single organs which may or may not develop.

It is impossible to foretell in a given case how many more symptoms will appear in the organism, or how severe a course the disease is destined to run. No favorable prognosis can be based on the mildness either of the initial symptoms at the point of infection during the primary period, or of the general symptoms in the skin and mucous membranes during the secondary period. Judging by our own experience, we can only say that strong, healthy individuals justify the hope that they will easily get over the disease, while patients who have been weakened by tuberculosis, malaria, or even by intercurrent diseases, will probably suffer more under similar conditions. It is therefore important to be even more cautious in prognosis with such individuals than in the case of otherwise healthy subjects. Very young, undeveloped individuals, and children who have become infected with syphilis, suffer more severely, as the tender, growing organism falls an easy prey to the ravages of the disease. The physician is therefore unable to say definitely to a patient that his malady will come to an end at such and such a time with such and such a symptom. As a rule, syphilis terminates with the irritative stage or the so-called secondary symptoms. Unfortunately, however, in many cases gummatous neoplasms develop in spite of the most careful treatment and management, and these neoplasms are capable of damaging the organism more than the milder inflammatory and infiltrative processes of the secondary period. In spite of the efforts of many skilful specialists, no definite signs have yet been discovered which enable the physician either to pronounce an attack of syphilis definitely ended, or justify him in expecting the occurrence of tertiary symptoms. Until new methods of examination are discovered which shall enable us to pronounce a body free from syphilis, we shall be forced to rely on certain empirical facts to determine whether or not the disease is permanently cured. These facts are: a healthy, vigorous condition of the general system before syphilis is acquired; a certain regularity in the appearance of the symptoms; the effect of appropriate treatment; and, finally, freedom from any symptom of the disease for a number of years. The following *scheme* will be found to hold good for most cases:

The interval from the time of infection to the appearance of secondary symptoms is about *eight weeks.* This period includes the development of the initial induration —within the first three weeks—the involvement of the neighboring lymph-glands and sometimes of the lymphatics leading to them, and subjective disturbances, complained of by some patients before the appearance of the rash. The first rash disappears, and after about *three months*—that is, *six months* after infection—a syphilide usually develops in some of the mucous membranes. The subsequent course of the disease is characterized by the appearance, at irregular intervals, of localized rashes which yield to appropriate treatment. With the end of this period, which usually lasts from *one and a half to two years,* the disease itself usually terminates.

An exception to this scheme is formed by the more malignant cases, in which there are no intervals of freedom from the disease, tertiary and secondary symptoms occur together, and nervous disturbances and general systemic disease manifest themselves early. These constitute the *malignant forms* of syphilis. Between these two varieties, in point of frequency, are those cases in which tertiary forms develop after a latent interval, during which the patient is entirely free from all symptoms.

THE PRIMARY STAGE OF ACQUIRED SYPHILIS.

Conditions of Infection.

Certain conditions are necessary for the infection of an individual with syphilis. In the first place, he must be free from the disease; and, secondly, there must be some loss of epithelium or epidermis—in short, some sort of wound—on the surface of the body for the infection to take place.

Syphilis in attacking an organism renders it immune against subsequent infection for a long time. It is true, even such an immune individual may be infected by an ulcer; but it is only an apparent syphilitic infection, since the ulcerative process which results is always purely local and entails no further consequence to the organism.

The abrasion, which is probably the most important condition of a syphilitic infection, may either be effected at the time of exposure—for instance, during coitus; or it may have existed before infection took place—for instance, an erosion after herpes præputialis or labialis. Syphilitic infection may, it is true, take place even if the tissues are not injured, but only if they are in a certain condition and the secretion which contains the syphilitic virus acts for a long period of time, setting up an irritation in the ducts of glands or in the delicate mucous membranes (preputial sac, rima pudendi); while on the other hand, even very short contact with an open wound is followed by infection.

Whether the infection occur through direct contact with a syphilitic body (*immediate infection*) or through some object polluted with the secretion of a syphilitic sore (*mediate infection*), the subsequent course of the disease is the same.

Channels of Infection.

An individual in the acute—that is, the completed, infectious—stage of syphilis is justly considered a menace to his surroundings. The virus is most abundant in the secretions of syphilitic ulcers, but it is also present in the blood and lymph during these stages. It is not contained in the normal secretions of the body, as the saliva, milk, and seminal fluid, although even these may become mixed with the virus on their way through the organism or at the points where they reach the surface, the oral cavity, the mamma, and in disease of the very vascular testicles which may not be demonstrable clinically. Persons with acquired as well as with hereditary syphilis, as, for instance, children with syphilitic pemphigus and papillomata, are a source of danger to a non-syphilitic organism for many years. Recent observations have shown that old sores which scarcely inconvenience the patient may give the infection long afterward, if they happen to become raw through maceration or some mechanical means. We may mention in this connection old anal and perianal infiltrations, tongue affections, and many other so-called places of predilection of the syphilitic products which will be discussed later.

It is generally held that the tertiary syphilitic products, the so-called gummatous neoplasms, cannot carry the infection, but the statement must be taken with a reservation. Thus, cases in which tertiary coexist with secondary symptoms, and which therefore belong to the group of tertiary forms, unquestionably do carry the infection. Still, the infective power of gummatous processes is undeniably weaker than that of secondary lesions, the principal reason probably being that the gumma disintegrates much later, after the infiltration has undergone a retrogressive metamorphosis, so that nothing is left but a detritus in which the syphilitic virus has become weakened or even destroyed. Another reason is that the gummatous forms are often localized in situations from which infection is impossible (internal organs), and, finally, the patients themselves dread and avoid contact with the gummatous ulcers on account of the pain it gives them. The secondary lesions, on the other hand, which lie more superficially and disintegrate very soon after they are formed, result in extensive tissue-destruction with copious secretion which is much more likely to carry the infection than that of gummatous ulcers.

Old ulcers resulting from gummatous wheals and tissue-alterations, like hyperostoses and eburnations in the bones, are to be regarded as the remains of old syphilitic products in those situations and contain no virus.

So far we have mentioned only human syphilitic bodies as carriers of the virus, but, in addition, a great variety of objects, such as spoons, glasses, wind-instruments, surgical instruments, bandages, etc., are capable of carrying the infection if they have previously come in contact with syphilitic wounds and some of the secretion has stuck to them. Desiccation of the syphilitic secretion does not render the virus harmless. Very high temperatures or freezing, on the other hand, so far as our present experience goes, are capable of destroying the virus.

The First Phenomena that Appear after Syphilitic Infection.

The first phenomena that appear after infection possess no characteristic features on which to base an opinion as to the effectiveness of the infection. They comprise wounds which existed before the suspected infection, or of fresh lesions sustained at the time of infection, or of macerations of the epithelium or epidermis in the region of the above-mentioned folds or ducts of glands, usually about the genitalia. The patients rarely consult a physician at this period unless they are in the habit of observing themselves closely and are aware that the lesion is directly due to contact with a foreign body. Persons who know nothing about syphilis or have been infected indirectly are often very slow to seek medical assistance, sometimes not before secondary symptoms have developed throughout the entire body. It follows from what we have said that the physician is seldom in a position to say, shortly after the occurrence of infection, more than that the erosion, or skin-lesion, or ruptured vesicles (as, for instance, after herpes) have a more or less suspicious look; he might perhaps determine by an examination of the individual suspected of being the source of the infection, whether he has to deal with a syphilitic affection or not.

It makes a difference whether the infection is due to syphilitic products alone, or whether a purulent secretion from a venereal infectious sore was inoculated at the same time. In the former case we have a slowly developing ulcerative process, or rather a gradual infiltration in the affected area; while in the latter case the acute course of the venereal ulcer, its rapid disintegration and profuse purulent secretion conceal the signs of syphilitic infection so effectually that it is only after the ulcer has healed that the specific nature of the process is recognized. All wounds of this kind, even venereal infectious ulcers, can be completely cured by antiseptic treatment and careful management; but it does not necessarily follow that the entire morbid process is ended. Infiltration often develops after scar-formation, the neighboring groups of glands become swollen, constituting a true initial infection, which produces the same effects on the organism as if it had originated in a simple syphilitic ulcer.

As a rule, however, the small, insignificant abrasions are gradually converted into rounded ulcers, spotted on the surface, which give but a scanty secretion and cause little discomfort to the patient. In the course of the second, and especially the third, week the typical infiltration develops at the base and near the periphery; it takes on a spherical nodular form, is hard to the touch, and represents the so-called **induration** or **sclerosis** (Pl. 1). The degree of infiltration depends on the local nature of the tissue in which the infection takes place. Thus, for instance, we frequently see on the glans penis and at the vaginal orifice flat infiltrations, eroded on the surface (erosio superficialis sclerotica). On the surface of the body, especially where the skin is loosely attached to the underlying tissue, we find nodular, hard infiltrations, elevated above their surroundings, which attain their greatest extent in parts covered with hair and rich in glands (Plates 2 to 11). We have seen such an induration about the size of a half dollar on the chin, which closely resembled a neoplasm, and it is not so very long ago that such syphilitic indurations on the lips, breast, etc., were mistaken for epitheliomata. If such initial lesions are situated in parts much subject to traction, as, for instance, at the corners of the mouth, at the junction of palate and tongue, in the tonsils, the anus, etc., they take the form of ulcerative fissures, and often complicate the diagnosis of induration by the rapid disintegration of the infiltrate. Syphilitic tissue in general, particularly an extensive induration, rapidly undergoes gangrenous decay from pressure or traction, so that the induration disappears and a large ulcer is formed. In the same way inappropriate treatment, especially useless, unnecessary cauterization, may cause the infiltrate to break down, and lead to the formation of a large ulcerated surface.

From the point of infection the syphilitic virus penetrates into the body through the lymph- and blood-channels. The lymph-capillaries especially take up the greater part of the tissue-juices flowing back from the periphery, and, uniting to form larger vessels, carry it to the nearest glands. The returning blood-vessels, the veins, also no doubt take up the virus from the induration and carry it to the rest of the body; but in the case of the lymphatics we have clinical and histological proof. Whoever has had the opportunity to observe a large number of cases will remember many in which the tissue immediately surrounding the induration was swollen and almost as hard and unyielding as the induration itself, was edematous to the touch, and sent out hard, tough cords to some distance, easily traceable to the glands of the region,—for instance, along the dorsum of the penis when the induration was seated in the prepuce. These changes begin with a capillary or cord-like lymphangitis with nodular swellings. If the lymphangitis is very superficial, or the nodular swellings are extensive, superficial excoriations or even complete decay not infrequently result. Such nodular ulcers may appear like so many separate points of infection; sometimes a lymphangitis of this character, following upon the initial lesion, may give rise to extensive swelling and result in a high degree of deformity involving the entire genitalia, a very obstinate condition which has been designated **indurative edema** (Plates 5, 12). It usually yields to general treatment and disappears entirely upon the advent of secondary symptoms. This variety of indurative edema is to be distinguished from acute inflammatory edema, an acute process which may accompany any septic or ulcerative wound, and occasionally leads to abscess-formation in the course of the lymphatics, as in venereal ulcer (bubonulus Nisbethii).

Swelling of Lymph-glands.

A constant phenomenon in the subsequent course of a syphilitic infection is swelling of the lymph-glands nearest the point of infection, indurations of the genitalia of the inguinal glands, induration of the mouth and lips (Plates 9, 10), of the submaxillary and submental glands, etc. It is the outward expression of a widespread round-celled infiltration which now attacks the lymph-glands themselves, after running its course in the initial induration and the lymphatics leading from it. The swelling is usually moderate in simple syphilitic ulcers and causes little discomfort to the patient. But in degenerating ulcers and in cases of mixed infection the glandular irritation is much more intense. The swelling is often so great as to form tumors as large as the fist, which break down at various points and present a so-called strumous adenitis. It often occurs in individuals weakened by scrofula, tuberculosis, etc., even when the peripheral irritation is comparatively slight.

Phimosis and Paraphimosis.

There is one very frequent complication of the initial form in the male organ, especially in the prepuce and neck. In such cases the prepuce, owing to the infiltration surrounding the induration, becomes completely immovable and rigidly adherent to the glans (Pl. 12). Such **phimoses** develop even when the fore-skin was originally quite loose, and are sure to occur if even a slight constriction was present. The constant pressure leads to resolution and not rarely to gangrene of the indurated area. If the gangrene is not checked, the prepuce becomes perforated and the glans may slip through the opening thus formed. We have seen cases in which the gangrene had completely destroyed the prepuce and had even spread to the skin of the penis and of the scrotum, so that the corpora cavernosa as well as the testicle and its supplying vessels were exposed. If such an infiltrated prepuce, before it becomes quite immovable, is forcibly pushed back, reposition is rarely possible (**paraphimosis**). The prepuce becomes edematous; and the neck of the penis undergoes necrosis. The circulation in the glans and in the retracted prepuce is impeded and, if the condition persists for some time, the edema is replaced by a permanent inflammatory infiltration, reposition of the prepuce is no longer possible, and a permanent deformity of the penis results.

THE SECONDARY STAGE OF SYPHILIS.

Prodromal Symptoms during the Eruptive Period.

While the local symptoms are developing with more or less intensity, the virus penetrates into the system from the point of infection by way of the lymph- and blood-channels, without other changes than those we have mentioned manifesting themselves for some time, usually until the fifty-seventh day after infection. The cases are, however, not uniform by any means. In a large proportion (perhaps more than half) the presence of grave systemic disease betrays itself by certain subjective symptoms during the period when the disease is spreading through the body, without manifesting itself by any marked external changes. The patients complain of lassitude and depression many days before the breaking out of the exanthema; they are pale, with black rings under the eyes—in short, they have a distinct morbid appearance. At the same time they complain of pains localized in different portions of the body: headache, intercostal neuralgia, pain on pressure in the sternum, especially near the points of union with the costal cartilages, without any demonstrable swelling, tenderness in single joints or groups of muscles, etc. The latter are sometimes called rheumatoid pains and are often mistaken for incipient rheumatism. The patients also exhibit a certain unrest and abnormal excitability, which in some persons merely take the form of irritability and moodiness, but in others, who had before been quite well, bring on violent palpitations of the heart at the least exertion, such as going up-stairs. Such patients also suffer from insomnia, either without any direct cause, or in consequence of the pains which usually increase toward nightfall, so that the general health becomes impaired. The symptoms we have referred to may be present singly, or several at the same time. A few patients have a slight rise in temperature of from 0.5° to 1°, toward evening, but this is the exception (syphilitic fever).

After a variable interval, usually from a week to ten days, these symptoms gradually disappear, often without any treatment, and the exanthema develops, which brings us to the *so-called secondary period.*

It will be well to mention now, before taking up the other morbid products of the secondary period of syphilis, that as a rule there is *a gradual swelling of the palpable glands* (*general syphilitic glandular swelling*) in the most varied regions of the body, usually observed about the same time as the first appearance of the general symptoms. The supraclavicular, cervical, nuchal, retro-auricular, the axillary, cubital, and other glands are found, either all or only in groups, to be somewhat increased in size and considerably indurated. The condition is most marked in scrofulous and anemic or otherwise debilitated subjects.

Even in later stages of the disease the lymphatic system plays an important part, both independently and in connection with ulcerative processes in the skin. In the secondary period we also have enlargement of the spleen, which, however, is difficult to detect and does not occur in every case. We shall return to these pathological alterations in a later section.

The Syphilitic Exanthemata.

The syphilitic exanthemata of the secondary period are divided, according to their histological appearance and clinical course, into three groups—*macular*, *papular*, and *pustular*. In addition, so-called squamous forms occur in very rare cases as a result of the papular exanthemata; they are called squamous because they early show a tendency to superficial desquamation (Pl. 20).

The syphilitic exanthemata exhibit certain *general characteristics* regularly seen in all forms. The rash is usually distributed symmetrically over both sides of the body and follows the fibers of the skin, especially in copious eruptions. Thus, for instance, on the back the eruption is disposed in parallel rows running obliquely downward on both sides (Pl. 16). Again, all syphilitic exanthemata, whether they appear singly or in groups of greater or less extent, as in the advanced stages, have a round or elliptical form (Pl. 21). This property, which is also found in other forms of dermatitis, probably depends on the distribution of blood-vessels in the skin. Lastly, it must be borne in mind that the eruption is only exceptionally uniform, in by far the greater number of cases it is polymorphous (Pl. 16). We observe either a direct transition of single eruptions from one form to another (for instance, from the macular to the papular) or one variety developing within another—for instance, papular forms in a group of macular eruptions, papular in a pustular group, with varying characteristics (Pl. 18).

The Macular Syphilide.—We distinguish two forms

of macular syphilides, *syphilitic roseola* and the *large macular syphilide.*

Roseola, the true representative of the hyperemic stage, chiefly affects the trunk; the macules are brownish-red in color and about as large as a lentil or a pea; they are not as a rule raised above the level of the skin, disappear on pressure, and may vanish completely, without leaving a trace, in a few days (three to twelve). As subjective symptoms are entirely wanting, this form of syphilitic eruption almost always escapes the patient's notice (Pl. 13).

The **large macular syphilide** occurs later than roseola, often in combination with papular forms, in the genitalia, on the anus, in the mouth, etc. In the lower part of the body the color is livid, in the upper it is of a distinctly coppery hue. The individual lesions, which on account of their size are called maculæ majores to distinguish them from those of roseola, called simply maculæ, have received various names according to their shape and distribution—Maculæ majores figuratæ, Maculæ majores gyratæ, Maculæ majores annulares, etc. Their formation depends either on the coalescence of several adjacent pustules or on the disappearance of the redness in the center and increased redness in the periphery of single pustules, causing these rings to appear more distinctly.

Most large forms of syphilide are slightly raised above the level of the skin, and hence resemble various forms of polymorphous exudative erythema. They can easily be distinguished from the latter, however, by their longer duration, the entire absence of subjective symptoms, and by other accompanying symptoms.

The large form of syphilide is not a mere hyperemia of the skin, as Biesiadecki has shown, but depends on round-celled infiltration about the blood-vessels in the affected area; it is therefore the first indication of the infiltrations that are characteristic of the papular stage.

As has been stated, roseola disappears without leaving any appreciable changes in the skin. In the large form, on the other hand, we observe in rare instances a slight, barely noticeable desquamation of the epidermis in the affected areas after the disappearance of the eruption. More frequently the pigmentation disappears, so that the affected parts appear white and lead to the formation of *cutaneous leukoplasia* (Plates 14, 15, 16).

The Papular Syphilide.—The commonest of the syphilitic eruptions of the skin, the so-called syphilides, is the papular form. It is often the first exanthema to appear after general infection of the organism, and runs its course either alone or combined with the macular or the pustular form. The base of the papule originally consists of a round-celled infiltration in the papillary layer of the skin, the size and shape of the papules depending on the extent and bulk of the proliferated tissue. In general appearance it is the most variable form of syphilide. We may have nodules ranging in size from a millet-seed to a bean, or more flattened papules, sometimes as large as a five-cent piece, presenting slightly raised edges and a somewhat depressed center. In recent cases of syphilis the papules are scattered over the entire integument, while in cases of longer standing they appear localized in the genitalia, the anus, the palmar surfaces, the mucous membrane of the oral cavity, etc. An opinion as to the gravity of a particular case may be formed from the shape and size of the individual papules.

The **lenticular syphilide** appears in collections of red nodules on the trunk and on the extremities. After a relatively short duration—eight to fourteen days—the individual nodules begin to show a dirty white discoloration and desquamate. This form of papular syphilide usually leaves no permanent marks (Pl. 17).

The smaller, so-called **lichenoid syphilide** usually attacks scrofulous or tuberculous individuals. It is almost never distributed evenly over the surface of the body, but occurs in groups of ten to twenty nodules. The individual papules rarely show much hyperemia on their first appearance, and soon undergo a yellowish discoloration, suggesting the picture of scrofulous lichen, especially if the eruption is copious. This form often persists a long time in spite of the most energetic treatment, and when it finally does yield, the nodules at first desquamate on the surface, but in the end the scabs come away bodily and leave minute punctiform depressions in the skin (Plates 19, 20).

Another form is the **flat, glistening, papular syphilide** (*papulæ nitentes*), which is usually observed on the nose, in the nasolabial furrows, on the forehead—in short, on the face generally. The individual papules exhibit a pale-red, shining surface, moderately raised, sharp edges, and a slightly depressed center. With proper treatment the papules desquamate and disappear, usually without leaving any visible alterations in the skin (Pl. 25).

Orbicular papules (Pl. 21) constitute a late form which frequently occurs in relapse and affects the seats of predilection (genitalia, anal region, etc.), or in connection with diseased organs (eye). The individual papules appear in the form of larger or smaller rings, according to the duration, exhibiting a slight depression with brownish pigmentation in the center, and separated by raised edges from the normal epidermis. The exudation at the borders of the papules is sometimes so profuse that the epidermis is loosened and forms a dry crust about the papule. As the sores begin to heal, the edges gradually flatten out and the center slowly regains its normal color by desquamation.

In conclusion, we may mention the **grouped variety of papular syphilides** (tubercula cutanea [Ricord], papulæ cumulis coacervatæ). This nodular form of syphilide occurs only in the later stages of syphilis, often associated with bone- and joint-affections, occasionally with true serpiginous ulcers. The individual groups vary in size from a half dollar to the palm of the hand, and are made up of infiltrates as large as a pea, occupying the entire thickness of the skin and covered either with layers of dead epidermis or with a thick, dry crust. The skin between these raised nodules exhibits a dark-red or brown pigmentation. This form of papular syphilide may persist for many months. It finally disappears either by absorption and superficial desquamation, resulting merely in a shallow depression in the epidermis, or by ulceration and superficial crust-formation, leading to deep scar-formation and permanent alteration of the affected area (Plates 22, 23). This late form of nodular syphilide resembles the destructive forms of the gummatous stage in its mode of healing, and is sometimes classed as a superficial cutaneous gumma.

The Pustular Syphilide.—The pustular syphilides cause vastly more discomfort to the patient than do the macular and papular forms. They rarely occur as an initial eruption, being preceded by macular and papular forms in the great majority of cases. Papular and pustular rashes sometimes exist together. There are cases, however, where the pustular syphilide constitutes the initial eruption, and these cases deserve special attention because they represent a more acute and rapid form of syphilis, in the prognosis of which the physician should be exceedingly cautious.

An eruption of pustular syphilide is usually preceded by grave general symptoms. There are evening rises in temperature; the patient looks pale and weak, he exhibits a strange restlessness, and often complains of lassitude, headache, and pains in the limbs. In this stage of syphilis we often meet with disturbances which point to disease of internal organs—for instance, icterus, albumin in the urine, etc. (Pl. 18).

We distinguish several forms of pustular syphilide.

Very often the eruption is characterized during its development by copious seropurulent exudations, the epidermis is loosened, and the pustules take on the appearance of **vesicles**—the so-called **vesiculous syphilide** (variola syphilitica). Later, the epidermis, together with the contents of the vesicle, forms a dry scab which is cast off and exposes the papillary layer. In most cases a newly formed epidermis is seen under the scab.

To this variety belongs the **pustula minor**, also called **acneiform syphilide**; the individual nodules resemble a papule with a purulent vesicle in the center. The affected areas usually correspond to the openings of the hair-follicles and ducts of sebaceous glands. The vesicle is soon converted into a brown scab which covers the dome of the pustule (Pl. 27).

The most important representative of the group is the **pustula major.** This pustular syphilide occurs either alone or in combination with the acneiform variety, and is characterized by its size, rapid disintegration, and a tendency on the part of the individual pustules to coalesce. The patient complains of burning pain, which is produced by the syphilide itself and becomes greatly aggravated when the clothing or bed-linen sticks to the freely secreting pustules. The crusts are frequently torn off and replaced by deep ulcers with a dirty white floor, which gradually destroy the entire infiltrated papillary stratum.

If the pustule spreads out superficially instead of attacking the deeper tissues, we get the so-called **ecthyma-pustule** (Plates 28, 29).

Sometimes the crusts which cover the pustules grow upward. The exudation proceeds slowly and gradually; the secretion dries as fast as it is produced; the scab increases in thickness, and, as it extends peripherally at the same time by the melting of the tissues, new scabs are constantly added, and a formation resembling oyster-shells is produced, to which the term **syphilitic rupia** is applied (Plates 44, 45, 49).

The subsequent course is the same in all forms of pustules—they heal by cicatrization. The scars frequently exhibit hyperemia and infiltration for some time. At last the hyperemia disappears, and the scar atrophies and becomes loose, white, and glistening, with an encircling zone of brown pigmentation. The alteration is permanent and especially noticeable on parts of the body covered with hair, if the roots of the hairs have been destroyed.

In conclusion, we would mention a phenomenon which rarely accompanies pustular syphilides. The individual pustules are surrounded by an irregular, bright-red zone several millimeters in thickness and resembling the erythema of erysipelas. Whether this is caused by the breaking down of the tissue alone or by the generation of toxins, we shall have to leave undetermined. One point should be emphasized: whenever we have observed this phenomenon the patient was much reduced and presented the appearance which is commonly seen in grave febrile diseases.

Syphilides with Cutaneous Hemorrhages.

Hemorrhagic syphilides are divided into two classes. The first class includes cases in which syphilis is complicated with another disease—for instance, hemophilia, scorbutus. Here the hemorrhage is a symptom of the complication, showing itself in the blood-vessels already suffering from the effects of syphilis.

The cases forming the second class are less numerous. In these the formation of papular or pustular exanthemata is accompanied by hemorrhages in the affected areas from the start, and without the coexistence of other disease, so that the hemorrhages must be regarded as the expression of disease of the vessels due to syphilis as such. The blood should be examined to determine whether the hemorrhages might not be in part due to grave blood disease.

The fact that the occurrence of such forms always points to grave disease of the general organism must not be overlooked in the prognosis, whether we have to deal with a grave complication of syphilis or with a particularly malignant form of the disease itself.

In the first series of cases the complicating affection must, of course, receive suitable treatment, just as in the second class the ulcers which almost always appear must be treated by suitable local measures in addition to the general treatment.

We shall discuss this question more in detail when we come to speak of treatment.

Abnormal Color-changes.

In most syphilides there is a shifting of the pigmentation; that is to say, the pigment disappears in the sore itself and becomes increased around its periphery. This is particularly the case in parts naturally rich in pigment, as the nape of the neck and the genital region. Occasionally the entire surface of the body is thickly covered with pale, non-pigmented circular or oval spots. If they come under observation early enough, the center is seen to be reddish and the periphery white, while the immediate surroundings are darkly pigmented. Later the entire area becomes white and appears the more distinctly for the darker pigmentation of the surrounding parts. This so-called **syphilitic leukoplasia** (Plates 14, 15, 16) is a more valuable sign than any other, as it may represent the remains of a cutaneous syphilide of very long standing, and forms a diagnostic point of the highest importance in doubtful cases of diseased organs, such as retinitis or endarteritis.

Many syphilides, especially such as are accompanied by great hyperemia, possess a directly opposite property of depositing large masses of pigmentation, which, if the eruption continues for a long time, leave dark-brown spots usually on the dependent portions of the body, persisting long after the disappearance of all other symptoms.

Finally, we may mention the pigment-destruction which sometimes results from the disintegration of pustular syphilides, destroying the papillary layer and persisting for life as whitish, thin, atrophic scars (Pl. 24).

Diseases of the Hairy Scalp.

During the secondary period of syphilis a variety of seborrhea of the scalp often develops, differing in many essentials from ordinary seborrhea. Instead of increased secretion, with desquamation of the epidermis, there is diffuse infiltration of the papillary layer and of the hair-follicles. The epidermis comes off in scales; the hair loses its color and gloss, and is easily pulled out, or even comes out of its own accord. The resulting baldness is usually fairly uniform (diffuse alopecia). Sometimes the disease presents the type of a papular syphilide, the hair is massed in thick bunches, and the loss of hair is confined to sharply defined areas about as large as a bean (areolar alopecia). The hair may be restored in both diffuse and areolar syphilitic alopecia. Unless the condition has lasted too long, lanugo hairs grow in three or four months, and are later replaced by strong, healthy hairs (Pl. 26a, black and colored).

The loss of hair is, of course, most marked in the scalp, but it is to be remembered that analogous processes may also affect the eyebrows, eyelids, and more rarely the axillæ and pudenda.

Pustular syphilide is a more common affection of the scalp than the above-mentioned diseases. Infiltration appears around the hair-follicles or attacks the roots themselves, so that while the hairs are still held fast on the surface by the incrustation, they die and become loosened within the scalp itself. The hair loses its gloss and soon falls out in large bunches, bringing the scabs with it. Occasionally a pustular syphilide in the scalp takes on a peculiar appearance. Either by extension of a single sore, or by the coalescence of several, the pustules attain the size of a quarter or a half dollar. The base proliferates and forms a mulberry-shaped tumor which may attain the size of a pigeon's egg (frambesia syphilitica), and the surface is covered with a dirty brown scab. The proliferated tissues bleed at the slightest touch and cause the patient much pain; they are very refractory to treatment (Plates 44, 45). Scars form after the pustules heal, and bald spots remain corresponding in extent to the areas destroyed by the process.

Diseases affecting the Palms of the Hands, the Soles of the Feet, the Fingers, and the Toes.

Psoriasis syphilitica palmaris et plantaris is the commonest form of syphilitic disease in the palms of the hands and the soles of the feet. It is a papular syphilide, which, however, appears much later than the exanthemata on other parts of the integument and presents some essential differences in its course. As the epidermis under which the papules form is very thick, it is often four months after the infection before they appear on the surface—long enough for the entire course of a papular form on the rest of the integument. In some cases psoriasis palmaris et plantaris appears much later, even several years after infection, the patient meanwhile being entirely free from any morbid symptoms. Lastly, the disease in many cases offers an obstinate resistance to every kind of treatment and persists after gummatous processes and disease in individual organs have already put in their appearance.

The papules themselves consist of small nodules varying in size from the head of a pin to a pea, and covered with a tough, horny epidermis. Often the patients do not become aware of them until they invade the region of the phalanges, and thus produce pain either by direct pressure or when the patient grasps any hard substance. Sometimes they appear in the form of flat, livid spots with horny, dirty yellow epidermis in the center; the skin finally cracks and comes off in scales. In rare cases the papules attain a large size; more frequently a number of them are crowded closely together, or even coalesce. The skin presents the appearance of infiltration, the horny epidermis cracks, and in a short time, especially if the skin is hard and thick, painful fissures develop at the flexures which seriously interfere with the use of the hands and feet, or even render them entirely helpless.

If papules are formed between two fingers or between two toes, the epidermis rapidly becomes macerated, an open sore results, and the fingers or toes, as the case may be, become swollen and livid, and their proximal extremities cut by radiating ulcers. Later the entire hand or foot becomes swollen and exceedingly painful, and, if the proper treatment is not administered in time, grave lymphangitis may develop.

If the papular infiltration attacks the matrix or margins of the nails, their nutrition becomes seriously impaired and **syphilitic onychia** or **paronychia** results. The distal phalanx becomes more or less swollen; the nail itself very horny, dry, and brittle, resembling a claw in shape; gradually it separates from the underlying tissue, the color changes to a dirty yellow or brown, the nail is pushed more and more forward and finally cast off entirely. If, as frequently happens, the process is accompanied by suppuration at the margin, the patient suffers intense pain, and the afflicted member becomes useless, especially as several fingers or toes are usually affected at the same time or in rapid succession (Plates 30, 31a, 31b, 32).

The disease usually lasts several months. The nails grow again, as a rule, though it may be only after the lapse of six months or even a longer time.

Secondary Syphilitic Phenomena in the Genitalia and about the Anus.

In the secondary period the genital and anal regions are most frequently the seat of grave phenomena which claim our attentive consideration on account of the regularity with which they appear, their tendency to recur, the great danger of infection, the variety of different forms, and, lastly, on account of the important *rôle* which the resulting alterations play in the diagnosis of visceral disease in later stages.

It is generally assumed that the tissues for some distance around an initial lesion are completely impregnated with the syphilitic virus, and are therefore in a state of irritation which affords a fertile soil for the production of new forms. This is especially true of the genitals, where the irritation is increased by the secretions, by sweat, and by want of cleanliness. Before the general integument becomes diseased, papules frequently appear at the edges of the labia majora in the female (especially in chancre of the vaginal opening) and on the scrotum in the male. Many people pay no attention whatever to such phenomena, either because they feel no pain, or because they are naturally indolent and careless, or attribute them to some other cause. Hence, relatively larger papular eruptions develop in these regions than on the rest of the body; the surface soon becomes macerated and ulcers are formed, with the production of detritus, pus, or only a serolymphoid secretion, according to the kind of degeneration present. These materials contain the most virulent form of the poison and are the most frequent source of syphilitic infection. The closely packed papules are at first very shallow; they soon run together and produce extensive ulcers, speckled on the surface and moderately infiltrated at the base, exuding a scanty secretion.

Sometimes a diffuse infiltration surrounds the lymphatics in the affected area—for instance, the prepuce, the skin of the penis, the labia majora—and the above-mentioned *indurative edema* develops.

The ulceration in genital and anal syphilide is rarely deep seated. As a rule, after a short duration—four to six weeks—the papules begin to grow upward from the base and often attain the size of a mulberry or a hazelnut. The proliferated masses are densely packed, raw on the surface, and present the appearance of proliferating venereal condylomata acuminata (also called venereal papillomata). Multiplying syphilitic papules (**papulæ luxuriantes**) are distinguished from the latter by the enormous proliferation and infiltration of the base, which is slightly raised above the level of the skin, and the absence of the deep fissures which separate the individual venereal papillomata down to their bases. There are also some anatomical differences between the two processes, the luxuriating papular syphilide being characterized by an abundant round-celled infiltration in the papillary layer of the skin, while the papilloma multiplies more in the epidermis.

In the **perineum,** on the **nates** (Pl. 37), and in the glandular parts about the anus the appearance of papules is attended with the same conditions as in the genitalia. In the **anal region** peculiar formations sometimes develop on account of the anatomical relations of the parts. The folds become longer, hard, and infiltrated; the intervals between them are marked by deep fissures which penetrate into the aperture of the anus. These fissures either radiate from the center, or they may be placed crosswise, so that the infiltrated folds are partially loosened from their bases. The general appearance suggests a number of fresh nodules in process of formation. The condition is painful in itself, and becomes doubly so during defecation, so that even the most indolent and careless individuals are led to seek professional advice.

It remains to be said that such processes result in infiltrates which penetrate deep into the skin, and, in spite of the most energetic treatment, often become the seat of new ulcerations. They are found by experience to constitute the most frequent source of syphilitic infection. It is not rare to see papular eruptions suddenly appear in the genitalia and about the anus after many years, when no other symptoms are demonstrable in the rest of the body. Such an eruption frequently occurs in the course of pregnancy as a result of the venous stasis in the genitalia. In prostitutes we have frequently seen single glistening, and later weeping papules of this kind appear in this region, when absolutely no other signs could be demonstrated in the rest of the body (Plates 33 to 39).

Diseases of the Buccal Mucous Membrane.

The mucous membrane of the mouth is almost always involved to a greater or less degree in the processes of the secondary period, besides being often the seat of primary lesions as a result of direct infection (Plates 8 to 11).

Thus we often see papules on the mucous membrane of the **lips** and **cheeks,** especially if they are already in a state of irritation from sharp fragments of teeth, or from excessive use of tobacco or other irritants. These papules differ from those on the skin chiefly by their rapid ulceration. As we can readily understand, the diseased spot on the mucous membrane, poorly nourished through its base, soon becomes macerated; the epithelium becomes cloudy and of a pale, whitish color; as early as the second day the surface breaks down, and we have a shallow ulcer with infiltrated floor which bleeds very easily (Plates 40, 41a).

Even more frequently than on the lips and cheeks we find papules on the **pillars of the fauces,** the **tonsils,** and the **soft palate.** They may be so numerous on the isthmus and in the throat as to simulate the clinical picture of croup or diphtheria. The diagnosis, however, is not difficult, since the condition is never accompanied by rise in temperature, the course is slow, and is further distinguished by the presence of other symptoms in the body. Disease of the tonsils occasionally gives rise to more or less grave functional disturbances. For, if they are severely attacked by the morbid process, they become much enlarged, the isthmus is narrowed, the patient's voice becomes nasal, and he suffers from excessive salivary flow and particularly from dysphagia. Sometimes the crypts break down and produce deep ulcers in the tonsils.

If the papules invade parts of the mucous membrane much exposed to traction, as the lips, angles of the mouth, and the base of the tongue, deep, cleft-like wounds, penetrating below the mucous membrane, are very apt to develop and cause the patient much discomfort.

The **gums** are less frequently attacked by papules than the rest of the buccal mucous membrane. The gums appear swollen and infiltrated, and the ulceration at the edges of the gums often loosens the teeth in their sockets.

In advanced cases of syphilis in the secondary period we occasionally meet with infiltrations in the buccal mucous membrane which are remarkable for their disinclination to form ulcers. We have seen such a diffuse infiltration in the soft palate and uvula, which converted the soft, flexible pillars into a tough, glistening band of a dark-red color and elevated above the surrounding tissue. The infiltrations shrink and produce a distortion of the velum palati and a retraction of the uvula to one side or the other (Pl. 42a).

The **tongue** is very often the seat of secondary syphilitic disease, which presents itself in so many various forms that it well deserves our interest. As there is an undoubted relation between mechanical irritation of a part and localization of syphilis in it, we must not be surprised that the tongue rarely escapes in a syphilitic attack.

As early as the papular stage individual papillæ on the dorsum of the tongue become more prominent and form spots the size of a pea, covered with loose, whitish epithelium. Later the epithelium comes off and the spots are converted into glistening, flesh-colored patches, and, if the process of maceration and disintegration goes on, into dirty yellow ulcers, raised above the level of the skin. This is particularly apt to occur on the dorsum and at the edges of the tongue, which are often intensely irritated by sharp, decaying teeth or remnants of teeth; the condition is very common in smokers and drinkers, especially if the care of the mouth is neglected. Such ulcers, of course, interfere greatly with speaking and eating.

Next in order after disease of individual papillæ we have a form which attacks larger areas on the surface of the tongue; the affected areas are sharply circumscribed, glistening, and slightly infiltrated, with a tendency to form superficial fissures and sores.

In a third variety, circular portions of the mucous membrane, as large as a penny, are converted into dense masses, distinct from the muscle and raised above the level of the tongue. The surface is marked by irregular furrows; here and there single, hypertrophied papillæ of a whitish color project from the surface.

Another form, which is often overlooked, attacks the region of the circumvallate papillæ or the adenoid tissue at the back of the tongue. In addition to the enlarged and infiltrated papillæ themselves there are other irregular ulcers which may defy treatment of every kind for a long time. Healing is followed by contracted scars, often covering large areas at the base of the tongue (Pl. 41b).

THE TERTIARY STAGE OF SYPHILIS.

The late manifestations of syphilis mostly take the form of gummata, hence this stage of the disease is called the gummatous stage, or, to carry out the analogy with the secondary, the *tertiary* stage. If these processes in the organism assume a malignant type and great destruction of tissue ensues, with the additional complication of amyloid degeneration of internal organs, we have the condition of syphilitic cachexia, which Sigmund has designated the *fourth stage of syphilis.*

The majority of syphilitic patients are fortunate enough to see their disease end with the symptoms of the secondary stage. The cases are rare in which tertiary and secondary products are present at the same time. They constitute what we have already referred to as malignant or precocious syphilis. In these unfortunate individuals the pustula major often appears as the initial eruption; at the same time they are tormented by periosteal gummata (tophi), and before the end of six months destructive processes begin their work in the cavities of the mouth and nose. To make matters worse, the usual remedies fail to arrest the malignant process, so that the patient's life is often put in jeopardy.

On the other hand, it frequently happens that the patient feels perfectly well for years after the completion of the secondary period, and is then suddenly reminded of his half-forgotten trouble by a renewed outbreak of morbid symptoms. The duration of this *latent stage* or *intermission-period,* during which the patient feels comparatively well, varies greatly. In one case thirty-four years elapsed between the disappearance of the secondary, and the advent of the tertiary symptoms; other observers put the duration of the intermission-period at from forty to fifty years.

Within recent years many attempts have been made to ascertain why tertiary forms should appear at all in certain cases. Some attribute it to inadequate treatment or to the entire want of treatment during the secondary stage; others are of the opinion that a disposition to the development of gummata may be produced by privation, or by tuberculosis, malaria, and other diseases which tend to weaken the system. As yet, the controversy is still in the theoretical stage, and the physician will do well not to make any promises to the patient after the disappearance of the secondary symptoms.

Many authorities assert that, in addition to a general disposition, an immediate cause is necessary to produce gummatous processes—for instance, a blow or other injury to a bone sparingly covered with soft parts, protracted excitement, alcoholic abuse, in the case of nervous disease, etc. This view served as the basis for the theory of the connection between syphilis and irritation, although it cannot be said to hold true in every case.

Many physicians lay down the universal rule that the

secretions of tertiary products, since they are not adapted to the inoculation of syphilis, can never carry the syphilitic contagion. We have already referred to this question in the introduction, and we again insist that the proposition must be accepted with great caution, and that it certainly does not hold in acute cases.

The tertiary stage differs in many respects from the secondary, not only in the nature of the lesion itself (gumma), but also in the manner of its occurrence.

The gumma is not preceded by general symptoms. Very often the patients are completely taken by surprise, and only begin to feel pain after the lesion has actually appeared; the degree of pain and interference with movement depends on the duration and seat of the process, and may be very great. It is only in consequence of the pain and discomfort that the patients show emaciation and other signs of disease.

Gummatous processes are further distinguished from secondary ones by the absence of symmetry in their distribution or regularity in their order of appearance. They are usually found only on one side of the body-surface, or even in one particular spot, and they not only persist for a long time, but may even recur in the same place. Sometimes the skin and mucous membranes are the seat of the first appearance of the gumma; again, the bones or even internal organs. In severe cases the process may, however, attack different parts of the body at the same time.

Gummata really represent a kind of neoplasm consisting of granulation-tissue. The nodes are composed of an irregular accumulation of granulation-tissue, in which the cellular element predominates more or less and which is in process of conversion into connective tissue; the normal tissue is crowded out and disappears entirely, or it becomes involved in the degeneration to which the syphilitic product itself falls a victim. At one stage of their development the tumors possess an elastic consistency, whence the name "rubber-tumor;" tumors of longer standing may be more hard.

The **tendency to degenerate** which characterizes all syphilitic products is shared by the gumma to a high degree. It is seen even in relatively recent gummata; the process begins in the center of the node, destroying the structure of the tissue, while at the periphery some newly formed connective tissue, well supplied with blood-vessels, remains and gradually merges into the adjacent tissue.

This characteristic property indicates the subsequent fate of the tumors. Gummata of subcutaneous and submucous cellular tissue, and subperiosteal gummata, often undergo rapid *mucoid degeneration.* Gummata in the glandular organs, liver, testes, and in the brain or in the muscles undergo *fatty metamorphosis,* and we may find *dry caseous masses* enclosed in an area of newly formed connective tissue, as in a capsule, where it remains for years without undergoing any change.

The gummata appear as individual nodes of varying magnitude; not rarely, however, a fresh node develops at the periphery of a former infiltrate, so that we see some nodes undergoing ulceration while new ones are forming about their periphery (*serpiginous character*).

We may also have multiple nodes appearing at the same time, or following each other at short intervals, so as to form groups of gummata; as the densely crowded nodes degenerate, the tissue lying between them is destroyed, and elliptical or kidney-shaped tumors or ulcers are formed.

Gumma of the Skin and Subcutaneous Cellular Tissue, The Gummatous Syphilide.

Cutaneous gummata usually appear during the second year after infection, but may also occur after many years of comparative good health. The circumstances which we have referred to as predisposing the organism to tertiary forms have the same effect with regard to affections of the skin, and we must further bear in mind that the general integument of the body is more exposed to injury than are the other tissues and organs.

An interesting fact, which has often been observed, is that gummata appear in places which were the seat of syphilitic products during the first and second periods; this may be regarded as a local disposition due to the former presence of the virus in the tissues.

The gumma develops in the cutis or in the subcutaneous cellular tissue. The size of the nodes varies from a pea to a pigeon's egg or larger.

If the gumma is superficial, the upper layer of the skin becomes livid and shares directly in the further pathological changes of the gumma.

If the nodes are seated more deeply, in the subcutaneous tissue, the skin is not involved until later; as the node increases in size, the skin becomes infiltrated and more or less inflamed. In both cases the skin remains intact and recovers its normal color if the proper treatment is employed and the node is absorbed.

If regeneration is too slow and the infiltrate becomes softened, the skin over it, which has meanwhile become very thin, also degenerates and an ulcer is formed. According to the seat and size of the nodes, the ulcers are more or less superficial, and the edges overhanging or steep and abrupt.

At first, the floor of the ulcer is covered with necrotic tissue; soon, however, the scanty purulent secretion of the ulcer dries, and, with the extravasated blood, forms a dark-brown scab such as we have described in connection with pustulous ulcers of the skin. If properly treated, the ulcer soon cleanses itself and healthy granulation-tissue is formed. Scar-formation begins at the periphery of the wound, and a flat scar eventually remains. Gradually the rest of the infiltrate disappears; the scar, which was livid at first, becomes white, and there is little disfigurement.

If several nodes develop at once and undergo rapid disintegration, large sinuous ulcers result. If the process continues and a new infiltrate is formed at the periphery, the ulcer becomes flattened on one side, but extends its limits on the other by fresh decay of the infiltrate, and we thus get **serpiginous ulcers,** semicircular or reniform in shape, with scar-formation at the center and ulceration at the freshly infiltrated periphery. If degeneration is very rapid, either on account of the reduced condition of the patient or of unfavorable local conditions, the tissue-destruction is very great and the ulcers may attain the most alarming dimensions. Thus we have seen the skin of the entire lower surface of the thigh destroyed by serpiginous gummata.

Under unfavorable conditions the products of tertiary syphilis, even more than those of the primary and secondary stages, are liable to gangrene. A tightly fitting garment pressing on the gummatous infiltrate often suffices to produce gangrene in an entire group of gummata; general nutritive disturbances may bring about the same result. I remember a case in which a "starving cure" (dry rolls) caused the appearance of dry gangrenous scabs in eight different places on the body. After the scabs had come away, shallow wounds remained which showed scarcely a trace of the nodular character of the gumma.

As to the nature of the scabs which form over gummatous ulcers, it may be said that simple ulcers with moderate secretion are covered with a scab consisting of a single layer; if the secretion is more abundant and the ulcer more extensive, the scab consists of several strata resembling an oyster-shell, like those described in the pustular syphilide—the so-called **rupia** of older writers.

The gummatous processes of the skin and subcutaneous cellular tissue are not confined to these structures; they often penetrate more deeply and involve the underlying muscles, bones, and joints (see Pl. 55). Myositis, caries of the bones, and even necrosis not infrequently accompany an advanced gummatous process, so that the disease becomes more dangerous, and worse consequences result. The tissues may be injured directly by the spreading gummata, or the resulting scars may be so extensive as to produce deformities or even destroy the movability of the limbs. In some cases the scars swell and proliferate, forming shapeless wheals (keloids) which may be present on the skin side by side with gummatous sores.

Before leaving the subject of gummatous affections of the skin, of which there are such countless varieties, presenting ever-varying pictures, we would mention a massive infiltration, one of the later forms of syphilis, which we prefer to designate **diffuse hypertrophic syphiloma** or **syphilitic leontiasis** instead of syphilitic lupus. This form only appears long after the completion of the secondary stage. It develops slowly, and is distinguished from the simple gummatous forms by its persisting for a long time without undergoing any marked change. It occurs in the form of hard, plate-like infiltrates on the lips, nose, and tongue. The infiltrate occupies the entire thickness of the parts mentioned, which eventually become quite immovable. The surface shows slight ulceration in places, but the process never attains the same depth and lateral extension as in the case of gummata. In every instance energetic antisyphilitic treatment is followed by absorption and healing of the sores; but for years afterward the site of the disease is marked by a moderate thickening of the connective tissue.

Syphilis of the Motor Apparatus.

As we have already remarked, the gummatous disease of the skin and subcutaneous tissue occasionally spreads to the underlying bones and muscles; bones which have but a thin covering are chiefly affected, as the anterior aspect of the tibia, the cranium, sternum, clavicle, ulna, etc. (see Plates 52a, 55).

The skin is naturally thin over these bones, and, if gummata develop, the bones very soon become involved, the periosteum is destroyed almost as fast as the skin itself, and the bones are left entirely exposed. If the proper treatment is employed, the patients may escape with a slight granular exfoliation of the bone; if, however, the tissue-destruction is extensive, and large portions of the bone are exposed, caries usually sets in and the bone is more or less completely destroyed. The bones which we have enumerated—including, perhaps, the ribs—often become the seat of spontaneous periostitis. Painful, slightly raised patches appear over the bone and increase steadily in size; unless the process is arrested by the proper treatment, the skin becomes inflamed, and soft ulcers (tophi) are formed which rupture toward the surface and discharge a mucous secretion.

Gummata proceeding from the skin, and involving the periosteum secondarily, find their analogue in a similar affection of the mucous membrane and the thin bones lying beneath it. The palate and the septum of the nose are chiefly affected; the periosteum is destroyed by the ulceration in a few days, and the bones are laid bare and fall victims to caries and necrosis.

So-called fibrous gummata frequently spring from the **periosteum**; these gummatous tumors do not undergo the retrogressive metamorphosis and rapid decay which we have described; they are hard and dense in structure, and embedded in a depression in the surface of the bone as in a niche. They yield to appropriate treatment by undergoing absorption, but they constitute a serious disease on account of their origin, duration, and the amount of destruction they cause in the bone. The surrounding portions of the bone become thickened and sclerosed. Such slightly raised hyperplasiæ are seen after periosteal processes of long standing in the flat bones of the skull, the anterior aspect of the tibia, etc.

Another form of gummatous disease of the bones, both long and flat, is **osteomyelitis**; in the long bones the process starts in the marrow, in the flat bones from the spongy substance, or from the diploë in the case of the skull.

The patients complain of boring pain, which usually comes on at night, a long time before any enlargement of the bone is noticeable. I am reminded in this connection of a very instructive case of pneumonia, in which gummata were found in several of the long bones at the autopsy (**osteomyelitis gummosa**).

The disease can undoubtedly exist for some time without producing any noticeable alterations, until finally the dense shell of the bone becomes enlarged or a central necrosis develops. The bones of syphilitic patients sometimes show a tendency to fracture from the most trivial causes; such cases are usually characterized by great shortening of the bone and disinclination to unite (*spontaneous fracture in gummatous osteomyelitis*).

In all such cases, whether they originate in a periostitis or in an osteomyelitis, if the bone is macerated, the center is found to be rarefied, while the substance in the periphery is increased in density.

If large portions of the periosteum are destroyed, or if several gummata exist side by side in the marrow, so that a large part of the bone is deprived of its nutrition, necrosis sets in and the affected portions of the bone are cast off as sequestra. If the soft parts become inflamed and ulcerate, caries also results.

Smaller long bones like the clavicle and the phalanges become rarefied by extensive gummatous infiltrations and produce the condition known as **spina ventosa.** It has been observed in the clavicle and also in the phalanges after syphilitic dactylitis.

Joints.—The synovial membranes of joints suffer in the same way as the periosteum, especially in the painful swellings of the joints which often occur in the early stages of syphilis (**syphilitic arthromeningitis**). Several large joints may become swollen, presenting the picture of articular rheumatism, and the diagnosis may be obscured by the exudation into the cavities of the joint, by the pain, and by the fever which is occasionally present. The condition is distinguished from rheumatism by the remittent type of the fever, the accompanying phenomena in the skin and mucous membranes, and by the shorter duration, especially if antisyphilitic measures are employed.

The prognosis in these acute forms of synovitis is favorable; if they are neglected, however, or if a cold-water treatment is resorted to, ankylosis results; there may even be crepitation, showing that erosion has begun in the cartilaginous investment.

But the syphilitic affections of the tertiary period are much more important than the disease we have just mentioned; grave alterations in the affected joint are almost invariably the result. We refer to gummatous disease of the bones or of the epiphyses, which have involved the joint, and peri-articular gummatous processes involving the fibrous capsules and ligaments, which have extended to the synovial membranes. Such articular cavities contain little serous exudation and are filled with adhesions, villous excrescences, and partially detached fragments of gummatous synovial membrane.

The usual outcome in grave cases of syphilitic joint-disease is **fibrous ankylosis,** even if the syphilitic process is peri-articular. More rarely peri-articular gummata rupture toward the surface, and continue to ulcerate until they break through to the articular cavity and set up a purulent articular inflammation. Speedy surgical interference becomes necessary in such cases, as antisyphilitic treatment is found to be useless.

Muscles.—The muscles also appear to be attacked early in the course of syphilis by rheumatic pains. But both the primary and secondary forms usually disappear of their own accord and leave no permanent consequences.

When gummatous disease of the skin and subcutaneous tissue penetrates to the muscles, the condition is more serious.

The muscle may also become the primary seat of a gummatous infiltration (**myositis gummosa**). The muscular gummata may attain a considerable size, as large as a hen's egg, and the course and location are sometimes such as to cause them to be mistaken for tumors, especially sarcomas. The disease begins with a round-celled infiltration, starting in the perimysium and those layers of the connective tissue which still contain blood-vessels; large portions of the perimysium become involved, and the transverse marking of the muscle-substance itself is gradually lost.

If such a gumma ulcerates, the muscle-substance may undergo necrosis and decay. As a rule, however, muscular gummata undergo fatty degeneration, and a cheesy mass of rather dense connective tissue becomes encapsuled. After the necrosed tissue has been cast off or absorbed, an extensive, fibrous scar, composed of the connective tissue which surrounds the gummata in large masses, remains, which destroys the function of any muscles that may be still intact, so that the extremities invariably become disabled (Pl. 55).

The infiltrative process frequently involves the **tendons,** especially the point of union with the muscle. All the tendons in the body are liable to the disease; we have observed it particularly in the tendo Achillis and in the ligamentum patellæ.

The **sheaths of the tendons** are also occasionally the seat of an extensive gummatous hyperplasia (Pl. 52). After a long time the infiltrate finally comes away spontaneously, but it is best to assist the process by surgical interference.

Syphilis of the Lymphatic Apparatus.

This heading includes diseases of the lymphatic glands, the tonsils, the follicles in the isthmus and in the throat, the spleen, the thyroid gland, and the suprarenal capsules.

The characteristic *glandular swelling* of the primary stage finds its analogue in the general glandular enlargement which occurs when the entire organism has become infected. It often makes its appearance before the skin-symptoms, but is sure to become aggravated if there is secondary ulceration in the skin and mucous membranes. In general, the glands are found to be more enlarged in scrofulous individuals and in those who have been weakened by disease. Swollen lymph-glands are often found in parts of the body (as the side of the thorax) where no glands can be felt normally. In the secondary period we can usually feel the inguinal glands, the cervical glands—from the mastoid process along the sternomastoid as far as the supraclavicular fossa—the axillary glands under the anterior margin of the pectoralis, the glands of the elbow over the internal condyle, etc. In autopsies on syphilitic subjects we have also found the internal glands swollen. The enlargement develops slowly and gives rise to elongated, spindle-shaped, hard nodes, although sometimes the spherical shape is retained. The glands shrink to their minimum with proper treatment. At first the glands appear reddish-brown in cross-section; later, the hilus becomes filled with connective tissue and sometimes with large masses of adipose tissue, so that the cortical substance of the gland is much reduced in thickness.

We have frequently observed glandular enlargement in the tertiary period. It is spontaneous and attains the size of a pigeon's or a hen's egg; it disappears if potassium iodid is given, and recurs either in the same or in other groups of glands. Glandular disease sometimes coexists with gummatous swellings in the skin; if the gummata are undergoing suppuration, they frequently give rise to glandular swelling (Pl. 54). Glandular swelling is not always directly associated with cutaneous gummata, and must therefore be regarded as a special disease, since we often observe gummata in the skin and subcutaneous tissue without any glandular swelling whatever.

These gummatous processes in the glands may lead to caseous degeneration, and the cheesy masses may remain encapsuled for a long time. The process may, however, start in the glands and involve the skin secondarily, producing inflammation and finally suppuration and necrosis; a typical case of this kind is shown in Pl. 53. In time the gland itself undergoes exfoliation (Pl. 54), and a scar results.

Spleen.—The spleen does not necessarily become swollen during the acute stage in every case of syphilis. But in a large number of cases, especially those in which the eruptive stage is complicated with chloranemia, a splenic enlargement can be demonstrated by palpation and percussion. It disappears when antisyphilitic remedies are given, just like the exanthemata of the secondary period. More rarely the organ remains indurated, the splenic pulp becoming harder and dryer; the connective tissue of the trabeculæ and capsule becomes thickened, and the latter may be attached to neighboring tissues by adhesions. These splenic tumors mostly occur in conjunction with disease of the liver, stomach, intestine, and kidneys.

Gummatous neoplasms have been observed as nodes of varying size in the interior of the organ, or more frequently beneath the capsule. They usually undergo fatty or cheesy degeneration, and can probably remain encapsuled within the spleen as a dry mass for a long time.

This circumscribed form of splenitis cannot be distinguished during life from the indurative diffuse variety which has been described. Anatomically such fatty or cheesy foci closely resemble infarcts; it is often difficult to distinguish them from caseous, solitary tubercles. Sometimes a similar localized degeneration of the parenchyma is produced by a syphilitic endarteritis.

The spleen is more frequently the seat of amyloid degeneration than other internal organs in those who have died of syphilitic marasmus; often we find amyloid disease in the spleen alone, the other organs being entirely free, or showing only traces of it.

Gummata have also been observed in the **thyroid gland** and in the **suprarenal bodies**; they are, however, extremely rare, and have so far only been found accidentally.

Syphilis of the Digestive Tract.

Oral Cavity.—We have seen that the mucous membrane of the oral cavity is almost always involved during the secondary period. Papules, ulcers, and fissures are constantly found. The alterations which are produced in the tertiary stage are known as *syphilitic pachydermia*, or *psoriasis mucosæ oris;* they occur in the mucous membrane of the tongue, the cheeks, especially opposite the teeth, and in several other localities. The characteristic sign is a thickening of the mucous membrane, with the formation of whitish patches, consisting of several layers of proliferated epithelium almost as hard and horny as epidermis. Other irritants besides syphilis have a share in the production of these patches, such as mechanical irritation by rough projections, sharp or decayed teeth, tobacco-chewing, smoking, and alcohol. The condition is incurable, and is very distressing to the patient on account of the tendency to form open sores; the patches are extremely vulnerable, and possess absolutely no elasticity, so that a morsel of hard food suffices to make an abrasion (Plates 41b, 42b). Rarely, submucous gummata form under these whitish, epithelial layers; much more frequently epitheliomata develop.

Gummata in the oral cavity proceed from the submucosa, but they invade the mucous membrane so rapidly that it is very hard to determine whether they really sprang from the mucous membrane or from the submucosa. They are usually found on the tongue, the palate, the isthmus of the fauces, and the nasopharyngeal cavity.

The Tongue.—There is scarcely an organ in which syphilis deposits so many and such various pathological products as in the tongue. The later stages of secondary syphilis are often marked by papular eruptions and ulceration along the margin of the tongue and by extensive infiltration on the surface. Among the tertiary forms we count alterations of the surface, so-called *psoriasis or leukoplasia of the tongue.* We may also mention smooth atrophy of the root of the tongue (*atrophia lævis baseos linguæ*), which, like psoriasis, is a persistent alteration and assumes a diagnostic significance in doubtful cases of syphilitic disease of internal organs. The process must not be confounded with cicatricial formations in this region, to which we have referred among the secondary affections. The atrophy develops without the patient's knowledge, probably in consequence of the lymphatic apparatus becoming involved, and corresponds to similar conditions in other tissues in syphilitic disease, as, for instance, atrophy of the heart-muscle.[1]

The development of a gumma in the tongue is a more frequent event. The gumma starts in the submucosa, and rapidly destroys the mucous membrane; but it soon heals and leaves a scar, if properly treated. If a large gumma, or several smaller ones placed close together, develop in the submucous and muscular tissue, the tongue becomes greatly swollen, and, if the growth cannot be arrested, the swelling soon softens. The mucous membrane is destroyed and a reddish-brown mass is discharged. The cavity which remains is often quite deep, and shaped like a fissure; its floor of a whitish color and covered with necrotic tissue. The tumors or ulcers are exceedingly painful and often prevent the patients from chewing and speaking, so that they soon become reduced in weight and strength.

If the gummata persist a long time, or recur frequently, they may give rise to epitheliomata. It is often difficult, on account of the cachectic appearance of the patient, to decide if he is suffering from a gumma or from a neoplasm. But carcinoma can be distinguished from a gumma by the lancinating pain, the condition of the glands, and, finally, by the failure of antisyphilitic remedies, so that the diagnosis is soon cleared up (Pl. 57).

[1] See Nothnagel's *Pathology;* Kraus's *Diseases of the Oral Cavity.*

Gummata in the tongue leave cicatricial contractions which may interfere materially with the use of the member and lead to injury and recurrence of the gumma, sometimes even to the development of an epithelial cancer.

The **base of the tongue,** the significance of which has already been referred to, is often the seat of gummatous neoplasms. They form in the adenoid tissue and produce ulcers and infiltrations which at first cause the patient no discomfort and therefore often escape detection. Unless the ulceration persists for a long time, the patients are not likely to have themselves examined. Palpation with the finger is as important in the diagnosis as inspection by means of the laryngeal mirror. The differential diagnosis from tuberculous ulcers and degenerated epitheliomata is undeniably difficult, and is based solely on the presence of other syphilitic signs and on the result of treatment.

The **junction of the hard with the soft palate** is a favorite seat of gummatous neoplasms, distinguished from all others by their rapid decay; before the patient has become aware of the disease, sometimes in a single night, a perforating ulcer develops. If the proper treatment is applied immediately, it may be possible to arrest the process and close the perforation, or at least to save a large part of the soft palate, so that the perforation can be closed by operative means after the ulceration has healed. But if the patient neglects to seek medical assistance, disintegration progresses rapidly, and after one or two weeks but a few shreds remain of the edge of the soft palate, from which hangs the infiltrated uvula. If these last remaining shreds tear through, the swollen uvula may be sucked into the air-tube and cause symptoms of asphyxia, so that it is best under such circumstances to remove it. The mildest result of such a destruction of the palate is a cicatricial distortion of the isthmus of the fauces; usually the arches are also involved in the degeneration.

Gummata in the hard palate undergo decay just as rapidly, whether they arise from the submucosa or from the periosteum, and a perforation soon follows. The bone itself is attacked by caries, and in a short time a large sequestrum is detached, or the bone gradually crumbles and the detritus is discharged. A communication is established between the oral cavity and the posterior nares, which causes the patient much distress even after the formation of a scar, as the taking of liquid and soft food, as well as speaking, becomes impossible. Some patients remedy the trouble by means of tampons, but it is better to close the opening with a rubber plate, or, if the loss of substance has not been too great, by operative means (Plates 56a, 56b).

Gummatous disease is sometimes primary in the **posterior wall of the pharynx,** starting in the pharyngeal tonsils, but it is more frequently secondary to disease in the nares or in the isthmus of the fauces. In an incredibly short time the mucous membrane is converted into a large ulcer by the rapid spread of the destructive process, and the pharyngeonasal cavity is occluded posteriorly; the destruction may extend to the periosteum and even to the turbinate bones (Pl. 56b).

The resulting deformities depend, of course, on the degree of destruction.

Even in the event of a cure certain deformities remain, depending upon the degree of tissue-destruction. The communication between the nose and throat is partially or completely cut off by cicatricial contraction of the remains of the soft palate and of the pillars of the fauces. This has the effect of drawing the base of the tongue against the posterior wall of the pharynx, although a small opening may still maintain some communication with the esophagus and the larynx. Swallowing becomes so difficult that operative measures must sometimes be resorted to. Respiration is also impaired when the nasopharyngeal cavity is closed, as the patient is forced to breathe through the nose. This gives rise to laryngeal troubles, bronchial catarrh, and deeper processes which interfere with the respiration even more seriously.

The **ear** is seriously affected by ulcerations in the pharynx; the orifices of the Eustachian tubes are destroyed, delicacy of hearing is lost, and the patient suffers severe stabbing-pains. If the disease attacks the middle ear, grave lesions of the organ of hearing may result.

The **mucous membrane of the cheeks and lips** and the **gums** are least frequently the seat of gummatous ulcers. If gummata and infiltrations do appear, they are usually situated on the lips, along the alveolar border (Pl. 41a). Such ulcers do not differ materially from those we have described. They may assume diagnostic importance in differentiation from tuberculous destruction of the mucous membrane or from epithelial cancer. In this connection it may be briefly mentioned that tuberculous nodules are very often seen at the periphery of tuberculous ulcers, and that the floor never possesses the enormous infiltration which is characteristic of syphilis; moreover, tubercular disease is rarely primary in this situation; we usually find at the same time advanced disease of the respiratory tract.

The course of epithelial cancer is, generally speaking, slower than that of syphilitic processes, especially gummatous infiltrations of the mucous membranes. In syphilis the submaxillary glands rarely become swollen, while they are always involved in cancer after it has persisted some time. Lastly, the diagnosis can be confirmed by means of the therapeutic measures which have been referred to.

The **salivary glands** have occasionally been observed to become diseased in syphilis.

Esophagus.—Syphilitic disease of the esophagus is usually diagnosed at the autopsy. The esophagus is never attacked primarily, but becomes secondarily involved in disease of the mediastinal glands and of the pharynx. Gummata in the mediastinal glands break through the wall of the esophagus, the mucous membrane is destroyed, and constricting scars result.

Stomach.—Acute or subacute gastric catarrh occurring in the early stage is very rarely the direct result of syphilis. It may possibly be considered so in cases where it forms a sequel to existing syphilitic disease of the liver or kidneys.

Ulcers consisting of gummatous infiltration of the submucosa occur in the stomach as the direct products of syphilis. They are often discovered at the autopsy, usually in the region of the pylorus and the lesser curvature, but occasionally also at the cardiac extremity; the infiltration develops in the submucosa and spreads to the mucous and also to the serous coats of the stomach. In addition to ulcerations, gummatous infiltrates and scar-formation have been found, so that it is fair to conclude that cicatrization of gummatous ulcers in the stomach is possible.

In rare instances we find ulcers due to syphilitic arteritis of the gastric vessels; they resemble the round gastric ulcer both in their clinical characters and in their anatomical appearance.

Intestine.—Acute intestinal catarrh, or chronic enteritis, occurring in constitutional syphilis, cannot be diagnosed *in vivo;* they not infrequently accompany syphilitic disease of the liver, such as amyloid degeneration, and therefore do not belong to the syphilitic process.

Ulcers, however, undoubtedly do occur in the intestine as the direct result of syphilis, but our knowledge of them is chiefly derived from accidental discoveries at the autopsy-table. They are usually multiple and localized in the small intestine, especially in the upper part. They usually develop from peculiarly rigid infiltrates, corresponding in position to the Peyer's patches, which penetrate the mucous membrane and the submucous and muscular layers, as large as, or larger than, a dollar. The mucous membrane is destroyed and irregular cavities remain, almost circular in shape and placed transversely to the axis of the gut, with punched-out edges of mucous membrane and a rigid floor, either covered with a grayish secretion or consisting of scar-tissue. The serous coat hypertrophies, and false membranes are formed which may occlude portions of the intestine. The scars which remain are flat and cause some stenosis.

In the **large intestine** the occurrence of grave disturbances is more common. Infiltrations of the anal folds and the fissures between them spread to the large intestine, or gummatous processes in the mucous membrane of the anus and in the perirectal tissues extend upward. The symptoms usually consist in the passing of mucopurulent matter, diarrhea, tenesmus, hemorrhages, and even prolapse of the diseased parts.

After a time the connective tissue increases greatly and pronounced constriction of the lumen results. This is a grave condition, and the patient rapidly becomes reduced by the intense pain, the fever, loss of blood, and excessive secretion in the bowel. Purulent periproctitis and even peritonitis may supervene and bring about the death of the patient.

The Liver.—Of all the internal organs the liver is the most frequent seat of the process. Whenever the internal organs are attacked by syphilis it is safe to assume that the liver is involved, even if the disturbances in other organs are the most prominent symptoms. The liver may also be the only organ diseased. Two forms are distinguished: *interstitial* and *gummatous hepatitis.* The two pathological alterations are nearly always associated; sometimes only portions of the organ are involved, sometimes the entire liver.

Hepatic gummata form larger or smaller nodes, usually about the size of a hazelnut, either single or disposed in groups so as to form tumors the size of a hen's egg. They are rather more common in the right lobe, and particularly affect the junction between the two lobes, under the suspensory ligament. They are usually found in a condition of necrosis or caseous degeneration, enclosed in strands of connective tissue of varying density which radiate more or less irregularly into the surrounding parenchyma, dividing the liver-substance into irregular islands. The proliferating gummatous granulation-tissue encroaches upon, and finally obliterates the liver-tissue, leaving only the smaller bile-ducts. The latter may be greatly hypertrophied. Gradually the cheesy masses are absorbed, the granulation-tissue partly disappears or becomes converted into fibrous connective tissue, deep, contracting scars appear on the surface, marking off whole sections of the organ—the *hepar lobatum.* If large numbers of gummata are massed in one situation or in one lobe, large areas or an entire lobe may disappear. If the gummatous process extends over the entire organ, there is a general increase in the connective tissue, subdividing the parenchyma into small islands—so-called *syphilitic cirrhosis.* It is characterized by an unequal distribution of the connective tissue, abundant at the seat of former gummata, less plentiful elsewhere. As the liver-tissue disappears, the parts that remain hypertrophy and regenerate, so that the islands of liver-tissue soon increase in size and produce lumpy excrescences on the surface, enlarged lobules on the cut surface, or enlargement of an entire lobe. Thus we have seen the left lobe hypertrophied to the normal size of the right, when the latter had become atrophied as a result of syphilitic disease.

In addition, peritoneal adhesions and distortions of the organ or of the gall-bladder and larger bile-ducts occur, and, with the interference to the portal circulation caused by the tumors and the overgrowth of connective tissue, add to the many disturbances which we observe in the living subject.

Pancreas.—Although the pancreas is very often diseased in the hereditary form of syphilis, the organ usually escapes in the acquired form in adults. Occasionally gummatous disease and alterations produced by disease of the vessels have been observed, but the cases are extremely rare, and the condition cannot be determined *in vivo* except by inference, when gummatous processes are present in other situations at the same time.

The Respiratory Tract.

The Nasal Cavity.—In the secondary period the mucous membrane of the nasal cavity is less commonly affected than is that of the oral cavity. The late lesions, on the contrary, are quite common and are of great practical significance. The most frequent seat of gummatous ulcers is the septum, especially the junction between the cartilaginous and bony portions. It scarcely needs to be said that the periosteum and investing mucous membrane rapidly break down and ulcerate. Before long the cartilage softens, the bone becomes carious, and perforation of the septum results. Such a perforation is sometimes discovered by the physician before the patient himself is aware of it. The secretion is purulent and mixed with blood, and has a most offensive odor; often it dries in the nose and forms brownish crusts which conceal the floor of the ulcer and the necrotic portions of the bone. The necrosis spreads along the line of the bony septum to the hard palate and a perforation of the bony plate of the palate results. The adjacent bones, the superior maxillary, the ethmoid, the internal pterygoid plates of the sphenoid, and the lachrymal bones may also be involved.

While ulceration and suppuration are going on, the sense of smell is entirely lost, and is seldom regained even if a cure is effected. Destruction of the cartilaginous septum results in the saddle-nose; destruction of the vomer and of the nasal bones produces sinking of the entire nose. It is often possible to save the skin of the nose even when there has been extensive destruction within the nasal fossa, but in time it withers and shrinks to a shapeless stump, nor is any plastic operation possible. In badly neglected cases the entire nose, including the skin, may be destroyed.

From the nasal fossæ the ulceration spreads to the upper lip, the alæ of the nose, the lachrymal gland, the posterior nares, and even as far as the head of the pharynx.

Even after the process has healed, there is danger of the scars in the nasal cavity reopening, and constant care is necessary to prevent a recurrence of the disease.

In the **larynx** the mucous membrane early (in the secondary period) shows signs of the disease in the form of papules, ulcers, and vegetations. The appearance of gummata in this important organ in the tertiary period is of the most vital significance to the patient. They are counted among the dangerous forms of syphilis, as the rapid disintegration endangers the cartilages of the larynx, the epiglottis, and the muscles and vocal cords with the mucous membrane covering them. It is, therefore, of the greatest importance to warn the patient of his danger, so that he may subject himself to the necessary treatment as early as possible. In themselves the gummata do not differ from those found in other mucous membranes, being characterized by the same tendency to rapid degeneration.

Infiltrations and ulcerations in the **trachea** and in the **bronchi** often follow syphilis of the larynx. The process may also begin in the mediastinum and involve these structures secondarily. The gravity of the symptoms depends on the extent and depth of the ulcerative process; very troublesome after-effects may remain even after a cure is effected.

Lung-disease in acquired syphilis is one of the rarer occurrences, especially if we exclude the cases which result from obstinate disease of the larynx and trachea. Cicatricial hyperplasia of lung-tissue and peribronchitis, associated with gummata, have been observed at autopsies. I have never been able to convince myself of the correctness of these anatomical changes by my own experience. Glandular tumors in the mediastinum may exert pressure on the bronchi and produce peribronchitis by contiguity. If the parietal layer of the pleura or the viscera are involved, we may have periostitis or necrosis of the ribs, and adhesions.

Syphilis of the Circulatory System.

Although the doctrine of syphilitic disease of the organs of circulation is one of the more recent achievements of morbid anatomy, more and more facts are being daily collected to show that it has received too little attention in the consideration of syphilitic products.

The **heart** presents many pathological alterations after syphilis. We distinguish three divisions: syphilitic changes in the smaller vessels of the heart, the products of syphilis in the pericardium and endocardium, and finally in the myocardium itself. Not to exceed unduly the limits of this sketch, we shall merely mention the most important alterations and the clinical phenomena to which they give rise.[1]

All the syphilitic affections of the heart that have been observed belong to the tertiary period; those observed in the secondary period have no anatomical foundation and must be regarded as functional disturbances.

Fibrous myocarditis is characterized by increase in the connective tissue and by the formation of wheals, and secondarily by atrophy and wasting of the muscle-substance. It is found, in limited areas, distributed over the interventricular septum and in the walls of the ventricles and auricles.

The gummata in the myocardium produce nodes of varying size, showing fatty or cheesy degeneration at the center and surrounded by layers of fibrous connective tissue. Like all gummata, they may remain in this condition for a long time, or they may break down and produce more or less destruction and loosening of the papillary muscles and valves.

Syphilitic endocarditis and pericarditis are usually but accompanying processes of disease of the myocardium.

In making a diagnosis *in vivo* it is to be remembered that the syphilitic symptoms appear most frequently

[1] For further details on "Syphilis of the Heart," see *Archiv. für Dermatologie und Syphilis*, 93.

between the ages of thirty and forty, associated with, or following upon the later lesions of syphilis. At a more advanced age disease of the heart is usually due to atheromatous change or to rheumatic endocarditis, or to fatty or fibroid degeneration, as, for instance, in chronic alcoholism, so that the diagnosis is extremely difficult. It is to be remembered also that grave functional disturbances of the heart may be simulated by syphilitic products in the central nervous system, as, for instance, in the bulb.

In disease of the heart-muscle manifesting itself in angina, palpitation, and severe dyspnea, it is easy to demonstrate the objective symptoms of dilatation, asystoly, arrhythmia, cyanosis, and slight anasarca. Other processes in the region of the valves may produce the symptoms of valvular insufficiency. Syphilitic endarteritis of one of the coronary arteries may give rise to the most intense symptoms of angina pectoris. For the benefit of the practitioner we repeat and emphasize the rule laid down by Semola: "If a patient, who has unquestionably had syphilis, presents himself with symptoms of a persistent arrhythmia, which refuse to yield to hygienic or pharmaceutical remedies, the physician must conclude that a syphilitic process exists, and must order specific treatment for the patient, even if at the time there are no symptoms which furnish ocular proof of the presence of constitutional syphilis."

Most cases of heart-syphilis are discovered accidentally at the autopsy. Death usually occurs rapidly and unexpectedly; rarely death is preceded by exhaustion of the degenerated myocardium with symptoms of cardiac weakness.

Although we are unable to diagnose syphilitic cardiac disease with certainty in the living subject, yet it is the duty of the physician, as stated in Semola's proposition quoted above, to fight the disease with iodids and even with mercury in doubtful cases, using the proper precautions, as the disease is always dangerous and almost certain to end fatally.

In **arteries** of medium caliber Heubner has described a specific affection which proceeds chiefly from the greatly hypertrophied intima; it is known as *endarteritis obliterans.* All the layers are affected; in some cases cheesy degeneration develops and destroys the adventitia and media, so that we are justified in speaking of *gummatous arteritis.* It occurs most frequently and in its most typical form in the arteries at the base of the brain; it has, however, also been observed in the carotid, the popliteal, the renal, and the splenic arteries, and in peripheral branches. The consequences are atrophy and necrosis of the organ (spleen); also encephalomalacia, to which we shall return later.

The etiology of *thickened valves* and *endo-aortitis* is more obscure, as these structures are so frequently the seat of general atheromatous changes; the same is true of the sequelæ, especially of aneurysms, and we take occasion to emphasize the fact that aneurysms never form in the cerebral cranial arteries in the typical form of the disease.

The smallest arterial branches are sometimes found to be involved in the late forms of syphilis; trophic disturbances and overgrowth of the connective tissue are the result.

The **veins** are rarely the seat of syphilitic products, although gummata have been described in the jugular and in the sheath of the femoral.

Syphilis of the Genito-urinary Apparatus.

Kidney.—The observations of the last few years have shown to what extent the kidneys may be affected by the syphilitic process. The presence of albumin in the kidneys of a syphilitic patient is not in itself enough to warrant a diagnosis of syphilitic disease of the kidneys. When mercury is given in the early stage of syphilis, the excretion of the mercury sets up an irritation in the kidney, and considerable quantities of albumin are found in the urine. Amyloid disease is found in the kidneys of cachectic individuals in the late stages of syphilis. These conditions make it difficult to determine whether the kidneys are directly concerned in the syphilitic process. Even in syphilitic subjects a nephritis based on diffuse interstitial proliferation is considered to be sufficiently accounted for on anatomical grounds. Still, gummatous tumors have been found accidentally at autopsies. It is not likely, however, that their presence would even be suspected on clinical grounds, unless perhaps antisyphilitic treatment were found to be followed by a good result in a case of violent renal symptoms pointing to degeneration of the parenchyma.

In the **bladder** ulcers have often been observed which were thought to be due to syphilis.

The **testicles** are much more frequently attacked by syphilis than is generally conceded. The disposition of this gland to specific disease may be explained on anatomical grounds, or by its liability to injury, or by previous disease. A slight infiltration, involving only the parenchyma, is not demonstrable, as the alteration produced is not severe enough to cause the patient to consult a doctor. In a few instances I have been able to determine an induration in parts of one or the other testicle in men who were given to observe themselves anxiously, or who, being on the point of marrying, wished to be examined for possible remains of syphilitic disease, or even were reminded of a former attack of the disease by the birth of children with hereditary syphilis. In two cases which I had known before the disease occurred I was able to demonstrate an induration in the tail (globus minor) of the epididymis.

Two forms are usually distinguished, *fibrous* and *gummatous orchitis.* Both belong to the late forms and develop two years or more after the infection.

Fibrous orchitis is more frequent and occurs earlier than the gummy form. It begins with an infiltration in the septum of the scrotum which spreads until it penetrates the parenchyma; the head (globus major) of the epididymis soon becomes involved. The patient feels no pain, only a sense of increased weight and traction in the testicle; the organ is found to be enlarged and harder than normal. If the process goes on, the body of the other testis is included in the tumor, which increases more and more in size. The patient complains of dragging pains as high up as the inguinal canal.

With potassium-iodid treatment the symptoms often disappear within a few days; but if the condition lasts, strands of connective tissue develop within the parenchyma of the organ which atrophies in parts and becomes permanently indurated.

Gummatous orchitis is distinguished by the appearance of a node, which gradually increases in size and becomes adherent to the investing membranes. The tumor is movable; if it breaks down, it bursts toward the exterior, and a brownish material, consisting of detritus and thin, watery pus, is discharged. In cases of long standing there is usually more than one node; the substance between the nodes atrophies from pressure, and what remains becomes hard from overgrowth of the connective tissue.

The tunica vaginalis propria also becomes involved in disease of the testis; the writer has often seen it hypertrophied and filled with a serous exudate. The scrotum may swell to the size of the fist or even the head of an infant; the tumor is excessively tense, so that it is impossible to distinguish the different parts of the organ by palpation. The closely adherent skin over a tumor of this kind may be destroyed in places. But, instead of a discharge of liquid products of degeneration, it is more common to have a solid mass of yellowish-white, fatty or cheesy material appear through the opening. If the process has gone as far as this, there is no hope of recovery, and amputation is the only relief.

The **penis,** especially the glans, the coronary sulcus, and the prepuce, is often the seat of a gummatous process which must be distinguished not only from nonsyphilitic tumors, but also from initial forms of syphilis (initial sclerosis). The gumma is much more prone to rapid decay than is the sclerosis; hence it is very important to recognize a gumma as early as possible so as to begin the proper treatment at once. The condition of the inguinal glands is of vital significance, since they do not become swollen in gummatous disease; in addition, anamnesis, duration, and course must be carefully taken into account. Gummata beginning in the coronary sulcus are very prone to spread to the glans and often destroy it in a very short time. A gumma situated near the urethra is dangerous on account of its tendency to penetrate into the corpus cavernosum.

Gummata in the skin of the penis are equally dangerous, not only because they destroy the skin and lead to external scar-formation, but chiefly because they spread to the corpus cavernosum. Occasionally the gumma starts in the corpus cavernosum itself. But in whatever manner the gumma may attack the corpus cavernosum, the infiltrate is sure to penetrate deeply and, after a cure by absorption has been effected, to leave cicatricial contractions. The consequence is that the corpus cavernosum is imperfectly filled with blood, and there is usually a bend in the penis at this spot during erection, or, if the destruction has been very extensive, erection is incomplete or even impossible.

In the **female genitals** the vulva, and in more advanced cases the vagina, are the most common seats of the process. Even under favorable conditions deformities, cicatricial contractions, and stenoses result; but if the process penetrates more deeply, it often produces perforations into the rectum. Usually a thorough and protracted course of treatment is necessary to arrest the ulceration, after which operative interference may be resorted to. If we at last succeed in arresting the process, the most we can hope to accomplish by an operation is to relieve to some extent the discomfort with which the condition is attended.

Uterus.—Syphilis may invade the uterus by direct infection through the os (Plates 6a, 6b, 7). The resulting sclerosis is followed by extensive infiltration and overgrowth of connective tissue at the cervix, which may under certain conditions interfere with parturition.

Papules may also appear on the mucous membrane of the vaginal orifice (Pl. 39). They are usually associated with papules on the external genitals and disappear with them under proper treatment. We may also mention the increase in size and density of the uterine tissue which is occasionally found in syphilitic parturient women. I have often seen labor interrupted by uterine hemorrhages, etc., in such cases, which usually ended in imperfect involution of the uterus.

Gummatous neoplasms have also been found in the uterine tissue, but they are no doubt extremely rare.

Gummatous processes occur in the **mammæ** in the later stages of syphilis; they usually proceed from the subcutaneous tissue and, by spreading to the mammary gland, produce the picture of a mastitis (Plates 48a, 48b). The skin over these infiltrations in the tissue of the gland is usually destroyed, and deeper parts may be lost if the proper treatment is not at once resorted to. I have known masses as large as a pigeon's egg to be destroyed in rapidly spreading infiltrations and serpiginous ulcers of the skin and subcutaneous cellular tissue in neglected cases. Potassium iodid is a valuable remedy in disease of this organ, as it is in disease of the testicle; its favorable influence on the process is manifest after a few days, and it may serve to clear up a doubtful diagnosis.

Syphilis of the Eye.

The specialist in venereal diseases sees principally diseases of the orbits, the eyelids, the cornea, the sclerotic, and the iris, as deeper-lying diseases of the organ and paralyses are usually taken to the oculist. We shall, therefore, in accordance with the plan we have adopted, confine ourselves to a short review of these affections.

Most forms of syphilis of the **orbit** occur as a periostitis, either localized in the orbital margin from the beginning or extending from the frontal bone to the orbital margin. We distinguish two forms: productive and destructive periostitis, both due to a gummatous process. The hypertrophic forms often begin in the secondary period, and are to be distinguished from hyperostoses of the orbits, as they are attended with considerable depositions on the orbital margin. The gumma attacks the skin as well as the bone, and may eventually perforate externally if it is not absorbed. The eyelid becomes edematous and remains immovable in a drooping position. If the levator palpebræ remains inactive for a long time, there is danger of the ptosis becoming permanent. If the destruction of the skin and eyelid is extensive, cicatricial contractions, ectropion, and lagophthalmos result.

The bones are more subject to gummy periostitis than to the hyperplastic form. A swelling in the periosteum and infiltration of the cellular tissue of the orbit may simulate an orbital tumor, as the same symptoms occur in both conditions. The upper wall is most frequently attacked, more rarely the thin, internal wall, the ethmoid bone. The disease is heralded by neuralgia and headache, becoming worse in the evening and at night; the pain is increased by touching the orbital margin. A characteristic symptom is dislocation of the globe, as it is a certain sign of infiltration in the periosteum and cellular tissue. If the periostitis is seated in front, the eyeball will be displaced laterally; if at the bottom of the orbit, it will protrude forward (protrusio bulbi). Syphilitic tumors are large, so that there is usually exophthalmos in addition to lateral displacement. A very characteristic phenomenon in syphilitic periostitis is interference with movement of the globe in one or more directions, even when the muscles are intact, showing that it is entirely due to dislocation of the globe. Eventually the muscles also become involved, and temporary or permanent loss of power results. If taken in time, the ulcers can often be cured; but if the infiltrate liquefies, the ulcer breaks through the anterior surface, usually at the orbital margin, even when the bone is still solid. If the periostitis runs on to necrosis, perforations into the nose and antrum of Highmore result; perforation into the cranium may produce a fatal meningitis.

The skin of the **eyelids** is often attacked in primary syphilitic disease. Later, papules may appear on the palpebral edges and on the conjunctivæ. Gummata produce a plate-like infiltration, and the conjunctiva assumes the appearance of trachoma from the newly formed granulations. Syphilitic tarsitis, an affection of the cartilage of the eyelid, has also been observed in the gummatous stage (Pl. 43b). Gummy processes in the lids may involve the conjunctiva and cause destruction or mutilation of the eyelid.

Diseases of the **cornea** belong to hereditary syphilis; they take the form of *interstitial* keratitis and are usually associated with diseases of the iris, the ciliary body, and the sclerotic. Gummata have been observed in the sclerotic, both primary and derived from the uveal tract, running on to degeneration and ulceration.

Iritis is frequent in the secondary, but very rare in the tertiary, stage of syphilis. Several forms are distinguished. The mildest variety is probably *serous iritis*. It is characterized by photophobia, moderate ciliary congestion, a slight discoloration of the iris, and the appearance of a deposit on the posterior wall of the cornea. The pupil reacts to light, and the deposit often has the form of a triangle with its apex pointing upward. In the second form, *plastic iritis*, the ciliary congestion is greater, the discoloration of the iris more pronounced, its markings indistinct, the tissue more spongy, pupillary reaction almost absent, and the pupillary margin of the anterior lens-capsule held fast by adhesions. The pupil may even be covered by a newly formed membrane (pseudomembrane). Deposits often appear on the posterior wall of the cornea. The aqueous humor is turbid, and on the floor of the anterior chamber there is sometimes found a hypopyon. In a third form, papular or condylomatous iritis, the ciliary margin of the iris is studded with small, miliary nodules, which spring from the tissue of the iris and have a reddish-yellow color (Pl. 43a). In this form the ciliary congestion and the posterior adhesions are more marked; the other symptoms are the same as those of plastic iritis. Unless relief is speedily obtained, various troublesome conditions may result—adhesions, occlusion of the pupil, and other grave disturbances.

We pass over the diseases of the ciliary body, the choroid, the vitreous body, the retina, and of the optic nerve, as they belong to the special province of ophthalmology both as regards diagnosis and treatment.

Syphilis of the Central Nervous System.

Syphilis of the Brain.—Syphilis of the central nervous system manifests itself in such a multiplicity of forms and gives rise to such a vast number of symptoms that it is difficult to distinguish it from other nervous diseases. There is hardly a symptom in the entire pathology of the nervous system that may not be produced by syphilis. A diagnosis of syphilitic nervous disease must therefore be based on a definite history, or on the existence of other processes combined with the nervous lesion that are positively known to be specific in character. Generally speaking, syphilitic nervous diseases belong to the late forms of syphilis, the majority appearing from the fifth to the tenth year after the disease has been acquired. In order to get a general idea of this extensive group of diseases we must bear in mind that the processes which eventually attack the nerve-substance, and either partially or wholly destroy it, may originate in various ways. Destructive processes in the bones extend to the dura and meninges and thence to the brain itself. Conversely, gummata in the dura produce pathological changes in the bone and in the meninges. Gummatous or chronic inflammatory processes may originate in the softer membranes, the arachnoid and the pia, and invade the brain secondarily. Lastly, arterial disease is the commonest cause of pathological alterations in the nerve-substance itself. The various symptoms which we observe consecutively in the living subject depend upon the seat and extent of the partial or total destruction of nervous tissue by the processes mentioned.

Syphilitic Diseases of the Periphery of the Brain.—Chronic, exudative, fibrous, hyperplastic inflammation of the meninges rarely occurs alone, being usually combined with gummata. The gummy depositions are found chiefly in the neighborhood of the cerebral arteries and nerves; the cranial nerves at their exit are enclosed in the masses of exudate in the subarachnoid space. The diffuse, fibrous neoplasms spread out over larger or smaller areas and produce adhesions of the meninges to each other and to the surface of the brain, so that several dry, cheesy gummatous foci may become enclosed. Fibrous as well as gummatous neoplasms extend to the cortex or even to the white matter. The cranial nerves which become involved in this process at the base of the brain are: the optic chiasm, oculomotor, trochlear, abducent, trigeminal, facial, glossopharyngeal, vagus, spinal accessory, and hypoglossal.

Gummata arising in the **brain-substance** itself often attain to a considerable size and, according to their seat, produce phenomena similar to those produced by other neoplasms. Such a gummatous encephalitis may be widely distributed over the cortex, the white matter, and the base of the brain without involving the meninges.

The **diseases of the cerebral arteries** are of the greatest importance; we have already made their acquaintance under the name of syphilitic endarteritis. They are more or less significant, according as the branches involved are end-vessels or not. The occlusion of the cerebral vessels which ensues is followed by nutritive disturbances in the brain-substance, softening, and destructive hemorrhages of greater or lesser intensity. Areas of softening or necrosis of the brain develop, such as we find, for instance, in the basal ganglia, the pons, and medulla oblongata, or there may be no more than a state of impaired nutrition, as there is a possibility of collateral circulation being established.

As we have intimated above, these anatomical alterations alluded to give rise to a very great variety of symptoms. The most important for the estimation of the gravity of the disease are headache, insomnia, vertigo, disturbance of consciousness and intelligence, etc.

Depending upon the position of the morbid focus, be it a gumma or an encephalomalacia, there will be paralyses and disturbances of sensibility.

The psychical disturbances which frequently accompany the conditions referred to have no definite anatomical foundation; they are caused by the grave nutritive disturbances in the brain either from the general cachexia or from a local process.

Diseases of the cortex may be produced by gummatous infiltrations or by encephalomalacial processes following the occlusion of a main artery (for instance, the middle cerebral artery), in which case the motor centers (the facial nerve), the centers of speech and of sensibility, as well as the intellectual centers, may be seriously injured. The paralysis may attack several groups of muscles in succession, or involve the entire extremity from the start. Sometimes cortical epilepsy occurs, showing itself in tonic and clonic spasms of single muscle-groups, or in epileptic convulsions affecting the entire half of the body. The epilepsy is attended with loss of consciousness. Temporary aphasia is usually present in cortical syphilis.

Cerebellum.—The diseases of the cerebrum which have been mentioned may under similar conditions affect the cerebellum also. In this situation they produce disturbances of the equilibrium, and are often combined with violent headache, vertigo, and pressure-sensations.

Although syphilitic processes in the brain present substantially the same symptoms as other cerebral diseases, there are certain combinations that are typical of specific disease and constitute a characteristic symptom-complex: this is the case in **syphilis of the base of the brain.** If the bones through which the cranial nerves pass are diseased, the nerves themselves become involved. The gummatous process attacks the perineurilemma and destroys the nerve-elements. Hence the chief consequence of disease of the base is the paralysis observed in these nerves, particularly the ocular nerves.

Gummatous diseases of the meninges which affect the pons or the peduncle of the cerebellum and produce paralysis on the side of the body opposite to the seat of the lesion are also accompanied by paralyses of the same side; this often happens in the case of the oculomotor, the abducent, and the facial.

Syphilitic disease of the arteries at the base of the brain, which we mentioned above, and gummata in the same situation are followed by very complicated phenomena, being characterized by paralyses in the peripheral distribution of the cranial nerves. In all cases of extensive gummata or foci of softening various nerve-paralyses are always associated with peripheral paralyses. It therefore becomes necessary to determine whether they are due to nerve-lesions within or without the brain-substance.

Syphilitic disease of the spinal cord, whether primary or secondary to disease of the meninges, is characterized by the occurrence of limited gummatous areas or by fibroid change in the meninges. Thus the myelomalacia may attack only peripheral portions of the white matter, or it may involve the gray matter as well. The resulting clinical pictures of myelomeningitis and myelitis present many variations, depending upon the particular segment or number of segments involved; we shall not enter into a detailed description of them at this point.

It might be well to call attention to a certain clinical

picture often seen in syphilis, a kind of tabes characterized, usually, by the absence of the Argyll-Robertson pupil and optical atrophy. Such cases should be carefully watched for some time and the effect of antisyphilitic treatment observed, before the patient is promised a cure. As regards the classic form of tabes, we know only that it often develops in syphilitic patients; we have as yet no knowledge of any intimate relation existing between the two processes.

The **nerve-roots** and the **cauda equina** may be attacked by secondary (meningitis), or by primary (oculomotor nerve) neuritis.

In conclusion, a word in regard to various neuroses in the peripheral nerves which we have observed in the course of a syphilitic process.

Such neuroses may occur in the distribution of the trigeminal or of the facial, or in peripheral nerves like the ulnar and sciatic. They begin with sensory disturbances which soon become extremely painful; later there may be motor and even trophic disturbances. The prognosis is good, however, and we have often been able to effect a complete cure by antisyphilitic treatment. But if the neurosis persists, atrophy and loss of function may result.

HEREDITARY SYPHILIS.

The term hereditary or congenital syphilis is applied to the disease when the fetal organism becomes infected *in utero*. The germ may become infected at the time of impregnation if the germ-cells of one or both parents are diseased (so-called *germinal infection*), or the offspring becomes infected during its development *in utero* before birth. While it is impossible to formulate definite laws of heredity, we submit the following generally accepted theories:

1. If both parents are syphilitic before conception, the more recent their own infection the greater the danger of infection to the offspring. The power of transmission in

most cases decreases from about the fourth year after infection, but it may be present as late as fourteen or fifteen years after that event. Syphilitic parents may produce syphilitic children, not only during the periods when demonstrable symptoms are present, but also during the intermission periods.

2. The most frequent infection is from the mother (*ovular infection*). Again, the duration of the mother's syphilis plays an important *rôle*. Recently infected mothers, with few exceptions, always infect their offspring; if, on the other hand, the infection is of longer standing, a relatively healthy child is occasionally born between two syphilitic children.

3. If the mother is infected during pregnancy, the child may also become infected *in utero* through the placenta. It is generally held that the placenta must first become diseased before such postconceptional infection can take place.

4. According to many observers, a syphilitic father is capable of infecting his offspring with syphilis at the time of impregnation (*spermatic infection*). If the mother has been infected at the same time by her syphilitic husband, she becomes immune through gestation of the syphilitic fetus; she can nurse her diseased child without incurring the danger of becoming infected by it (Colles's law). Exception has been taken to this law because once a mother was infected by her syphilitic child, and in another instance a mother who had given birth to a syphilitic child was infected at the end of her pregnancy.

The syphilis of the parents, especially of the mother, has such different effects on the syphilis of the offspring that little or nothing positive can be said concerning its influence.

If both parents, and especially the mother, contract syphilis shortly before conception takes place, the fetus soon dies *in utero* and abortion occurs during the third or fourth month. Various factors are concerned in this result: the diseased embryo is incapable of development, the mother's nutrition is so much impaired by her disease that she is unable to nourish the fetus properly, the uterus itself may be diseased, or the placenta is so altered by the syphilitic disease of the blood-vessels that the fetus cannot be nourished and therefore dies.

While this is unfortunately the most frequent termination of such pregnancies, there are other cases in which the fetus goes on developing until the seventh or eighth month and is then born prematurely, a sickly child, or is stillborn at the end of pregnancy. Finally, the child may be born alive, but with such grave syphilitic disease that it succumbs in a few hours or days. Such births occur chiefly when the parents have been suffering from syphilis for some time and in cases where the virulence of the disease has been reduced by palliative treatment.

Parents in the gummatous stage of syphilis usually produce healthy children. The author has known many cases of comparatively severe tertiary disease on the part of the mother during pregnancy, which nevertheless terminated favorably. It is a curious and hitherto unexplained fact that a relatively healthy child, practically free from specific symptoms, may be born between two diseased children.

It follows from what has been said that it is impossible, in a case of positive syphilis of the parents, to predict whether the offspring will be healthy or diseased; experience, however, teaches that careful treatment of the parents affords the best guarantee for the proper development of the offspring.

The history of many countries which have been visited by epidemics of syphilis shows what a decimating effect hereditary syphilis has on the population. To-day the eloquence of figures confirms what had been obtained by induction:

Von Wiederhofer estimates the mortality of syphilitic infants to be 99 per cent. Fournier has constructed the following table for the mortality of the offspring of syphilitic parents;

When both parents are syphilitic, 68.5 per cent.
When the mother is syphilitic, 60.0 per cent.
When the father is syphilitic, 28.0 per cent.

Hereditary syphilis is conveniently divided into an **early (syphilis hereditaria præcox)**, and a **late** form (**syphilis hereditaria tarda**).

A fetus that dies early rarely exhibits any pathological changes. In stillborn children certain pathological alterations are almost constantly found in many organs; the most important of these are osteochondritis at the epiphyses of the long bones, enlargement of liver and spleen, *abscess in the thymus gland*, and disease of the heart and blood-vessels and of the digestive tract. To these are added, in the order of decreasing frequency, diseases of the nervous tissues, of the kidneys, testicles, etc. It is not within our province to describe these important and interesting alterations in detail, and we shall merely mention a few phenomena of practical clinical importance observed in hereditary syphilis.

One absolutely fatal form of the disease sometimes declares itself at birth; the term **syphilis hæmorrhagica neonatorum** is applied to it. The principal alterations occur in the blood-vessels and give rise to hemorrhages into the parenchyma of organs, particularly into the cellular tissue in which the vessels lie embedded. Death occurs very early, in the first few hours or at most within two days, with the symptoms of cardiac weakness, cyanosis, sometimes with peripheral edema, anasarca, or even ascites. The many hemorrhages on the body resemble small petechiæ; sometimes more extensive extravasations are found.

The principal symptoms observed in syphilitic infants are the *nutritive disturbances* to which they are subjected while still *in utero*. The body-weight is usually below normal; the skin is withered and hangs in folds. Of the diseases which present visible alterations the most frequent is the so-called **syphilitic pemphigus**, which either exists at birth or develops within the first three or four days. The vesicles of this exanthema are collapsed and either burst or dry up, leaving excoriations or sores covered with scabs on the thin, pale skin (Pl. 58). The internal organs of such children are always diseased, and they pine away in spite of the most careful nursing. They usually die of inanition in one or two weeks; it is rarely possible to keep them alive longer than that.

In the very first days of the infant's life **disease of the navel** plays an important part; in spite of the greatest care hemorrhages and ulcerative processes occur and are often followed by general septic infection.

Septic infection and the entrance of various bacteria are favored not only by the umbilical lesion, but also by the many sores on the skin caused by the exanthemata. Furunculosis, long held to be a syphilitic product in the skin, is an example. All these adverse conditions combine to bring on the speedy death of the already debilitated infant.

Another very frequent disease is **snuffles,** which early declares itself in syphilitic infants. The mucous membrane of the nose becomes swollen; dyspnea is the most distressing symptom, the excessive secretion being of secondary importance. The infiltration spreads from the mucous membrane to the perichondrium and periosteum, and finally results in the formation of a *saddle-nose* (Pl. 60c).

Before long, *papular eruptions* appear on the nates and about the genitals at the junction between mucous membrane and skin. The papules become macerated and form ulcers or fissures at the angles of the mouth and about the nostrils. In addition, *vesicopustular exanthemata* appear on the skin which soon desiccate and leave brown scabs.

The matrix of the nails, the palms of the hands, and the soles of the feet are also attacked by eruptions.

Diseases of the eyes, such as iritis, keratitis, etc., occur; they give rise to pronounced swelling of the conjunctiva, purulent secretion from the conjunctival sac, and to erosions and fissures at the corners of the eyes.

These symptoms are associated with *enlargement of the liver and spleen, affections of the bowels, pulmonary catarrh*, which are caused by aspiration of the secretions from diseased buccal and nasal cavities, and not infrequently cause death within two or three months.

If the infant lives through all these graver troubles, other phenomena soon make their appearance. The **diseased epiphyses,** which hitherto had escaped notice, become detached, beginning usually at the humeral joint. At first a painful swelling appears about the affected epiphysis and the limb hangs down as if it were paralyzed. Sometimes it is possible to determine the condition by the abnormal movability, but usually the picture is simply that of palsy.

Even after all the symptoms in the skin, mucous membranes, and joints have disappeared, the children are still weak and anemic, and we can readily understand that they suffer more severely than other children from intercurrent diseases, such as bronchitis, pneumonia, intestinal catarrh, etc.

This early period of hereditary syphilis is sometimes followed by **late forms,** like the tertiary forms of acquired syphilis, for which the term **syphilis hereditaria tarda** is used.

It is asserted by some observers that this late form of syphilis may occur between the ages of sixteen and twenty or over, without having been preceded by any other form of the disease. But it is to be remembered that hereditary syphilis does not necessarily produce very marked changes in the child, and we cannot therefore agree with this view. Good health is a relative term and very elastic in its application. A slight nose-trouble and tardy development are phenomena which may easily escape detection. In treating advanced forms of syphilis I have found the marks of severe processes in the skin, glands, and joints which had not been attributed to syphilis. There is still a general impression that syphilis always gives rise to definite symptoms, especially such as can only be cured by specific treatment. These and many other reasons account for the different and erroneous views which are held about the late form of hereditary syphilis.

It is important to distinguish between **infantile acquired syphilis** and the hereditary form. I have often had occasion to treat patients between 16 and 18 years of age for the later forms of syphilis. In such cases the most careful inquiries must be made whether the parents ever suffered from diseases which might have had a specific origin, what the history of the confinements in the family has been, whether any children were stillborn, and whether there are any brothers or sisters affected with syphilis. If any such signs of hereditary syphilis were found, the patients never failed to show disturbances of development or other signs of hereditary disease. If no such indications were found, the case usually turned out to be a tertiary form, the infection having been acquired early, without the knowledge of the patient or of his family.

The late forms of hereditary syphilis begin to show themselves about the fifth year, sometimes with the appearance of puberty about the twelfth year; we have observed them to last, with interruptions, as late as the twentieth year.

It is not our purpose in this place to discuss in detail all the protean forms to which hereditary syphilis gives rise sooner or later in the organism; we shall content ourselves with mentioning the most striking symptoms which help the physician to determine whether a given morbid product is to be attributed to syphilis or not.

To Hutchinson is due the credit of collecting a group of symptoms which enable us to distinguish tertiary hereditary forms from acquired syphilis with great accuracy. They are: *A deformity of the permanent upper incisors*, consisting in a crescentic notch on the free edge of the teeth. *A dimness of the cornea*, or *an existing interstitial keratitis* (Plates 60a, 60b; black plate), and *rapidly increasing deafness.* If we add a prominence of the frontal protuberances, flat or even depressed nose, and fine scars at the angles of the mouth, the upper lip, and the mucous membrane of the lips, radiating from the nares, the picture is complete. Fournier is quite right when he points out the tardy development of such individuals. They look like children even at the age of 16 or 18; the genitals, the pubic hairs, the breasts in the female subjects, are imperfectly developed.

As to the **symptoms** of tardy syphilis, they include the lesions in the bones and joints which we find in the acquired form, and skin-lesions in the form of gummatous, serpiginous ulcers. The skin of the nose seems to be a favorite seat. This is perhaps due to the chronic catarrh which is so often present; ulcers gradually form, the process spreads to the cartilages and to the bones, and at last the external skin is also destroyed.

Certain groups of **lymph-glands**, notably the cervical glands, undergo a peculiar form of hyperplasia resulting in large tumors, which show no tendency to soften or break down. The condition usually lasts a long time and resists treatment much more stubbornly than do the other symptoms. Such glandular tumors resemble sarcoma.[1]

The **internal organs**, liver, spleen, and kidneys, are usually enlarged, and often irregularly contracted by gummata and cord-like neoplasms. The most frequent alteration in these organs is amyloid degeneration, especially if the patient is debilitated.

The **nervous system** does not escape the ravages of the morbid process. The patients frequently suffer with epileptiform attacks; they possess a low order of intelligence and are sometimes even half-witted. The same diseases of the meninges, the arteries, and of the brain-substance occur as are found in acquired syphilis, and the

[1] As in one of my cases in the Rudolfsspital; see Yearly Report, 1892.

diagnosis must be based on the evidences of the presence of syphilitic processes in other parts of the body and on the loss of the cerebral functions.

It is difficult even to outline the proper **treatment for hereditary syphilis.** In every case the patient requires the most careful management, if there is to be even a chance of success. Drugs should be administered from time to time, particularly iodids and tonics; in the early forms modified inunction cures and sublimated baths may be employed. The greatest benefit is derived from careful dieting, good, pure air, and baths containing iodids—more than from a general treatment.

THE TREATMENT OF SYPHILIS.

We shall divide the treatment of syphilis into three subdivisions: 1. Initial forms and their immediate consequences; 2. Local treatment of the secondary and tertiary lesions; 3. General treatment of syphilis.

1. Initial Forms and Their Immediate Consequences.

It is the duty of the physician to treat every suspicious lesion as if syphilitic infection had actually occurred. The ruling principle must be to destroy the syphilitic germ at the point of infection as soon as possible. As we have seen by the pathology, the first diagnosis can only be a provisional one, even if the case is seen in the first few days, but the physician is nevertheless justified in adopting energetic measures at once. It is impossible to say how long the virus remains at the point of entry; certainly a very short time, for I have never had any good results from excision of a simple initial sore within thirty-six or even twenty-four hours. Although these excisions were made before any evidences of reaction to the poison had made their appearance, the virus was not removed, showing that it had already been communicated to the vessels and other parts of the body outside of the initial sore. Nevertheless, excision is advisable in certain seats of syphilitic infection which are favorably situated for operation, as fissures in the preputial margin, in the edges of the labia, etc. The excision, to be justifiable, must be made within the first few hours after infection; I have never known it to be successful in cases where a beginning infiltration or even a complete sclerosis could be demonstrated, although the treatment of the initial lesion was somewhat simplified by the operation. What has been said about excision is equally true of the use of the thermocautery; the burn must, however, be treated longer than the wound from an excision, which may be allowed to heal by first intention. Cauterization is indicated only when the sclerosis has become phagedenic and is rapidly degenerating.

In the initial forms we confine ourselves as a rule to the *antiseptic treatment,* using the customary drugs, corrosive sublimate, carbolic acid, salicylic acid, etc. For many reasons, learnt in the prophylactic treatment of syphilis, it is not advisable to begin a general treatment at once, except in cases where the sclerosis is seated on the face—at the lips, for instance—and tends to degenerate rapidly, with the formation of large glandular tumors. In such cases the practice has always been to endeavor to stay the destructive process by immediate resort to general treatment.

In the **male genitals** the sclerosis is usually seated on the prepuce itself or in the neck of the penis, and is often associated with *phimosis.* If the foreskin was too long originally (præputium perlongum), and is converted into a hard mass by the infiltration spreading beyond the initial sclerosis, it must be removed with the knife in order to shorten the treatment and put an end to the patient's distress. It is not necessary, or even advisable, to remove the entire prepuce; it is enough to open the preputial sac at the dorsum and excise a portion of it.

If the sclerosis becomes *gangrenous* and destroys the inner layer of the foreskin or the neck of the penis, immediate operation becomes necessary. After the operation such phimoses heal readily and give no further trouble; even the remains of the edges of the sclerosis heal and leave a scar.

If there is a *paraphimosis* from forced retraction of the infiltrated prepuce, surgical interference becomes necessary to prevent further destruction of the prepuce and of the skin of the penis by the pressure of the constricting preputial ring.

The *dorsal lymphatics,* owing to the infiltration in simple, non-ulcerative scleroses, are converted into hard, sometimes nodular welts; these may be covered with strips of gray plaster; they soon disappear when the general treatment is begun later on.

The *glandular swellings* of syphilis rarely go on to suppuration unless they are complicated with venereal ulcers, or the patients are scrofulous or otherwise debilitated. In such cases the adenitis is to be treated surgically in the manner indicated in the section on venereal ulcers.

In treating scleroses in the **female genitals** the only difficulties are encountered in applying the remedy and getting the bandage to stay. Powders and strips of gray plaster are employed with advantage; a T-shaped bandage will keep the dressing on the external genitals in place.

Scleroses at the *orifice of the urethra* and *about the anus* may be treated at first with bougies and suppositories of iodoform (iodoformi puri, 0.1 (gr. xv); olei theobromæ, q. s. u. f. suppositor. urethrale), and later with strips of gray plaster. Sometimes a dressing of antiseptic gauze, saturated with a 5 per cent. solution of white precipitate or a 1 to 3 per cent. solution of red precipitate may be employed with advantage. The dressing must be changed from one to three times a day, according to the amount of secretion, and the ulcer cleansed each time with carbolic acid or corrosive sublimate.

2. Local Treatment of the Secondary and Tertiary Lesions.

In **syphilitic affections of the mouth, nose,** and **throat** the patient must use a mouth-wash and gargle several times a day, especially after meals and before retiring at night.

The following prescriptions are recommended:

℞. Pot. chlor., 10.0 (ʒijss);
Aq. dest., 500.0 (Oj).
S.: Gargarisma.

℞. Pot. chlor., 10.0 (ʒijss);
Alum. crud., 1.0 (gr. xv);
Aq. dest., 400.0 (f℥xiij);
Aq. menth. pip., 100.0 (f℥iij).
S.: Gargarisma.

℞. Acid. borac., 10.0 (ʒijss);
Solve in aq. dest., 500.0 (Oj).
S.: For external use.

℞. Acid. salicyl., 3.0 (gr. xlv);
Spirit. vin. rectif., 30.0 (f℥j);
Aq. dest., 300.0 (f℥x);
Tinct. krameriæ, gtt. xxx;
Pot. hypermang., 5.0 (ʒjss);
Aq. dest., 100.0 (f℥iij).
S.: One tablespoonful to a glass of water for mouth-wash and gargle.

℞. Tinct. cascarillæ,
Tinct. cinchonæ,
Tinct. krameriæ,
Spirit. menth. pip., *āā*. 25.0 (ʒvj).
S.: Twenty drops to a glass of water for mouth-wash.

Ulcerative, papular sores on the **mucous membrane of the lips** and **cheeks** and on the **isthmus**

of the fauces must be touched with nitrate of silver every day by the physician himself (solid nitrate or a 5 per cent. solution). If the patient is a responsible person, he may be trusted with a sublimate mouth-wash, either in the form of tablets prepared with NaCl, 1 gram (gr. xv) each, to be dissolved in water, and used in the proportion of one tablespoonful of the solution to ten tablespoonfuls of water as a gargle; or

℞. Hydr. bichlor. corros., 1.0 (gr. xv);
Alcohol. absol., 50.0 (f℥jss).
S.: Poison.—One tablespoonful to a glass of water for mouth-wash.

These sublimate gargles are to be used two or three times a day, an entire glassful being used each time.

Ulcers and fissures on the tongue must also be touched with nitrate of silver. Hard infiltrates on the surface of the tongue should be painted with a more concentrated alcoholic solution of sublimate, or with tincture of iodin, once or twice a day.

Destructive gummata and ulcers on the mucous membranes must be energetically treated with solid nitrate of silver and thoroughly cleansed with one of the abovementioned mouth-washes or with an irrigator. In using an irrigator for the nasal cavity, care must be had not to exert too much pressure, lest the liquid be forced into the cranial cavity or into the Eustachian tube and cause pain. If there are any necrotic bone-fragments at the bottom of the ulcers or between the proliferations of granulation-tissue, they must be removed as soon after they have separated as possible. If the operation is attended with hemorrhage, which is often the case, absorbent cotton or adhesive iodoform gauze may be used to control it.

The local treatment of **laryngeal syphilis** requires some proficiency in the use of the laryngeal mirror and other instruments, and should only be attempted by a specialist. *Inhalation* of the vapors of iodid solutions (2 per cent. potassium iodid), as an adjunct to the general treatment, is allowable in very light cases; it is, however, usually inadequate for the removal of the larger forms of papules, or of extensive infiltrations, or even of deep gummy ulcers.

For **papular syphilides on the genitalia and anus,** the most frequent form of syphilis in the female, and very apt to recur in both sexes, we use Labarraque's dressing:

℞. Chlorin. liquid., 20.0 (f℥v);
Aq. dest., 80.0 (f℥ijss).
S.: Apply with a brush.

And

℞. Calomel,
Amyl, *āā*.
S.: Dusting-powder.

The papules are first moistened with the chlorin-water and then dusted with calomel. Sublimate in the nascent state is thus formed, which acts as an intense caustic without giving much pain.

In the severe **hyperplastic forms** affecting the labia majora, the perineum, and the nates, the infiltration may be made to subside rapidly by painting daily with a stronger alcoholic or ethereal solution of sublimate (1 : 20) and covering the moistened parts with strips of cotton. The use of caustics, such as sublimate collodium (1 : 20), Plenck's solution [1], and others, has been given up as too dangerous as well as painful. Even the solution we have indicated must be applied with great care, so as not to touch any but the hyperplastic tissues; if the burning is severe, it should be followed by the application of aluminum acetate or Burow's solution in the form of com-

[1] ℞. Hydr. chlor. corros.,
Aluminis, *āā*. ℥j;
Plumb. acetat.,
Camphor., *āā*. ʒj;
Alcohol.,
Acid. acet., *āā*. ℥xij.

presses. Inunctions with stronger white precipitate sometimes have a good effect.

℞. Hydrarg. ammoniat., 5.0 (ʒjss);
Unguent. emollient., 40.0 (ʒx);

either alone or reinforced by 0.1 (gr. jss) sublimate.

For flat papules which do not secrete much a good adhesive gray plaster answers every purpose.

Fissures about the anus, which are usually found *close to proliferated anal folds,* often defy all remedies and are best treated surgically. The proliferations are removed with Paquelin's thermocautery, the patient being anesthetized, and the wound is afterward dressed with iodoform vaselin or with white-precipitate ointment. Tying the infiltrated folds with elastic ligatures is not to be recommended, although I formerly employed that method in a good many cases without injury to the patients.

Palmar and plantar psoriasis, fissures and degenerating papules between the toes and on the fingers. In all these forms the hands and feet should be softened by soaking in warm water (with soap) and then dressed with good, soft gray plaster. Deep fissures or ulcers with signs of inflammation are to be treated with baths and compresses dipped in Burow's solution. Sometimes it is advisable to apply sublimate solutions (1 : 1000) in addition to the bathing. A very good plan is to rub the affected parts with the above-mentioned white-precipitate ointment (4 : 40, with 0.1 to 0.2 (gr. jss to iij) sublimate) after bathing them at night, and then to put on a pair of Swedish leather gloves.

Syphilitic onychia and **paronychia** are treated with warm-water baths, compresses, and washing with sublimate, the diseased end-phalanges being well tied up in caps of gray plaster. If the edges of the nails are turned in, a local anesthetic is administered, the nail is split, and the edges are cut away. A protecting bandage or finger-cot should be worn constantly until the nail grows again.

In diseases of the scalp, which we have described, some form of local treatment is always necessary. First of all, the hair is to be cut short. The patient must wash his head every day with soap and water and rub it with white-precipitate ointment (1 : 10). If pustular ulcers or deep, destructive gummata are present, the scalp must be dressed with iodoform vaselin, mercurial ointments, or with gray plaster, and the dressing held in place by a properly applied head-bandage. Carious bone-lamellæ either separate of their own accord or, if necessary, are removed with forceps. The local treatment must never be omitted, as it may be the means of avoiding an extensive necrosis of the bone even when large areas are exposed. I have seen holes in the skull as large as a dollar heal over and form a scar. If the loss of bone-substance is considerable, insert plates of celluloid after the scar has formed, in order to protect the brain.

The treatment for the **various cutaneous processes**, the **muscle-, bone-,** and **joint-lesions** of the later stages of syphilis, cannot well be reduced to a schedule. If the numerous external remedies are insufficient to control the process, and surgical interference is indicated, it should not be delayed too long. We are getting over our reluctance to operate on syphilitic patients.

3. General Treatment of Syphilis.

To begin with, the so-called *expectant treatment*, by which is meant the regulation of the diet and other hygienic conditions, without the use of the specifics mercury and iodin for the local lesions, is in our opinion entirely inadequate.

As soon as syphilis produces lasting functional disturbances in the circulatory organs, neuralgias or pathological changes in the skin or mucous membranes remote from the point of infection, the physician must begin a specific mercurial treatment and persist in it until the morbid symptoms have disappeared completely. In fact, experienced practitioners make it a rule to continue the treatment for a period equal to at least a third or even a half of the duration of the symptoms, after the latter have disappeared.

Any subsequent relapse, if accompanied by marked alterations in various parts of the body, must again be subjected to both general and local treatment. Thus we follow the pathological process and keep the patient under observation, treating him with mercury, or later with iodid, only when he actually shows unmistakable syphilitic symptoms. We do not advocate the practice of giving drugs at definite intervals, whether the symptoms are present or not (chronic intermittent treatment —Fournier), because we have noticed that such a course does not prevent the recurrence of the symptoms after a certain period; besides, the patient becomes accustomed to the drug and does not respond so readily when it becomes necessary to give it to him at the next outbreak of the disease. But this is not the place to discuss this question in detail; we only wished to define our position.

For headache, insomnia, and other distressing symptoms which sometimes occur during the eruptive period, we prescribe:

℞. Pot. brom.,
Sod. brom., āā. 4.0 (ʒj);
Ammon. brom., 2.0 (ʒss).—M.
Ft. pulv. No. x.
S.: 1 or 2 powders at bedtime.

Or

℞. Pot. brom.,
Pot. iod., āā. 0.5 (ʒjss).—M.
Ft. pulv. No. x.
S.: 1 or 2 powders at bedtime.

Before the appearance of the exanthemata and in the intervals between the relapses (if there are any) and the consequent specific treatment, we have various duties toward the patient to fulfil. We must *prepare him for the various forms of mercurial treatment*—the mouth and the skin must be looked after, the initial lesion treated locally; and after the course of specific treatment is over he requires a supplementary one, such as medicinal baths, tonics, etc.

THE CARE OF THE MOUTH BEFORE AND DURING MERCURIAL TREATMENT.

We cannot emphasize too strongly that a systematic, careful treatment and supervision of the mouth is an absolute necessity during the entire course of mercurial treatment, and even longer. The patient must clean his teeth three or four times a day with brush and powder. In our hospital practice we use the following:

℞. Pulv. dentifr. nigr., 50.00 (℥jss).

Or

℞. Calcar. carbon. pulv., 50.00 (℥jss);
Magnes. carb., 10.00 (ʒijss);
Pulv. rad. ir. flor. (orris), 20.00 (ʒv);
Olei menth. pip., (gtt. lx).—M.
S.: Tooth-powder.

After that the gums are painted with an astringent tincture, and then rinsed with water.

℞. Tinct. iodin.,
Tinct. gall., *āā*. 10.00 (ʒijss).
S.: Apply with brush.

Or

℞. Tinct. krameriæ,
Tinct. gall., *āā*. 20.00 (ʒv);
Olei menth. pip., (gtt. xl).
S.: Gum tincture.

Or

℞. Ol. cadin.,
Spir. vin., *āā*. 10.00 (ʒijss);
Tinct. laud. simpl., 5.00 (ʒjss).

As a gargle we prescribe hypermanganate of potassium in the usual pink solution, 0.5 to 2 per cent. potassium chlorid, in solutions of 1 to 2 per cent. with the addition of alum, or

℞. Aq. calcis,
Aq. destill., *āā.*

Or

℞. Acidi salicyl., 5.00 (ʒjss);
Spir. frument.,
Aq. destill., *āā.* 100.00 (f℥iij);
Ol. menth., (gtt. v).—M.
S.: One teaspoonful to a glass of water.

In **mercurial stomatitis**, in addition to the abovementioned local measures, we paint the gums with nitrate of silver (1 : 30 to 1 : 15), rinsing the mouth with a saline solution, or with a 15 to 25 per cent. watery solution of chromic acid. Erosions and ulcers are touched with the solid stick of silver nitrate, or the recently suggested combination of silver nitrate with a 25 per cent. solution of chromic acid may be employed. By this method the newly formed argentic chromate leaves a red scab on the surface of the ulcer.

Mercurial Treatment of Syphilis.

The sovereign method of applying a mercurial cure is **inunction with gray mercurial ointment**; it is applicable not only to all forms of syphilis, but to all ages as well.

The infant can be rubbed by the mother or nurse after the morning bath with dilute mercurial ointment (1.0 (gr. xv) unguent. ciner. with 1.0 (gr. xv) ung. simpl.) on the sides of the thorax and on the abdomen. An adult weighing about 60 kg. should use from 4 to 5 g. (ʒj to ʒjss) daily. The inunction ought to be performed by an expert masseur, but if that is impossible, as in hospital practice, the patient must be taught to do it himself. A small amount of the ointment is taken at a time and rubbed in alternately on both sides of the body till the skin is quite dry. The places selected for the inunction are usually the calves of the leg, the inner surfaces of the thighs, the epigastric region, the lateral regions of the thorax, the inner surface of the upper arm and forearm, and the back. After all these parts have been well rubbed, the patient should take a warm bath for the sake of cleanliness.

The inunctions are to be kept up until all the symptoms have disappeared, and after that for an additional period equal to one-third or one-half of the time occupied by the original treatment. If the patient is unwell from any accidental cause, or if he suddenly develops pronounced mercurial salivation and stomatitis, the treatment must be discontinued for a few days until the troublesome symptoms have subsided, and then resumed. There is no objection to repeating the treatment for every one of the many relapses which usually mark the course of a syphilitic disease.

In addition to the inunctions, other measures, such as the baths we have referred to, and internal remedies, tonics, etc., may be employed.

The patient must be well fed and housed while he is undergoing the treatment. If his work exposes him to hardship and fatigue, especially to the inclemencies of the weather, the treatment should not be used. We have never seen any good result from it under such circumstances; the patient becomes anemic and the treatment seems to do him more harm than good, so that he only loses faith in it.

The physician must give his attention to the care of the mouth, as in all forms of mercurial treatment, and also to the digestive organs and to the excretions. We have on several occasions found albumin in the urine, and have been obliged to reduce the dose or to interrupt the treatment; in a few cases we have even been compelled to give it up altogether.

After the specific treatment is completed, the patient's mode of life must be properly regulated; a mild water-cure, invigorating baths (salt-water baths), or, if possible, removal to a milder climate, should be recommended to supplement the specific cure. The mouth needs careful attention for weeks after the specific treatment is completed.

For the inunctions we prescribe in the case of adults:

℞. Ung., ciner., 4.0–5.0 (ʒj to ʒjss);
Dent. tal. dos. ad chart. cerat. No. v.
(Dispense five such doses in waxed papers.)
S.: To be rubbed in (as indicated above).

According to the Austrian Pharmacopea, 7th edition, the ointment is prepared as follows:

℞. Hydrarg.,
Lanolin, *āā.* 200.0 (℥ij),

to be rubbed in until it disappears and then followed by the application of:

℞. Ung. simpl., 200.0 (℥ij).

For **extensive cutaneous syphilids, multiple ulcerative processes on the surface of the body, cutaneous exanthemata of the skin,** and for **weeping papules in small children,** sublimate baths should be employed. 10–15 g. (ʒijss–ʒjv) in tablets. For adults 10–15 g. (ʒijss–ʒjv) of sublimate solution are added to a bath of from three to five buckets of water at 26° to 28° R. (32.5° to 35° C.); the patient remains in the tub from ten to fifteen minutes, during which time he is gently rubbed. After the bath he must stay in bed. Young children are simply sponged with the solution once or twice a day; the diseased portions of the skin are first carefully washed and then dusted with a suitable powder (pure starch or starch mixed with calomel).

Hypodermatic Methods.—(*a*) **Soluble Mercurial Preparations:**

℞. Hydrarg. bichlor. corros.,	0.10 (gr. jss);
Sodii chlorati depurati,	1.00 (gr. xv);
Aq. destill.,	10.00 (f℥ijss).

Inject one syringeful of this solution daily, either subcutaneously in the back, or between the muscles (intramuscular injection) in the nates.

The following solution has been employed with great success, especially in the treatment of "walking cases":

℞. Hydrarg. bichlor. corros.,	0.5 (gr. viij);
Sodii chlorati,	2.00 (℥ss);
Aq. destill.,	10.00 (f℥ijss).

Inject one syringeful of this solution at intervals of five days or a week. If the precaution is taken to disinfect thoroughly both the syringe and the place where the point is to be inserted, and the injection is followed by gentle massage of the part, there is practically no danger that the wound will be painful or that an abscess will form.

(*b*) **Insoluble Mercurial Preparations.**—The use of these preparations is to be avoided, as a rule, for they all have the disadvantage that the amount of mercury absorbed necessarily remains an unknown quantity, as absorption varies greatly according to circumstances which cannot be controlled. Grave accidents sometimes occur in consequence of the sudden absorption of large quantities of the material supplied.

However, we give some of the more common formulæ:

℞. Hydrarg. salicyl.,	1.00 (gr. xv);
Olei oliv. optim.,	10.00 (℥ijss).—M.

S.: To be injected.

Shake before using.

℞. Hydrarg. oxid. flav., 1.00 (gr. xv);
Olei oliv. optim., 10.00 (ʒijss).—M.
S.: To be injected.
Shake before using.

℞. Calomelan., 1.00 (gr. xv);
Olei oliv. optim., 10.00 (ʒijss).—M.
S.: To be injected.
Shake before using.

Internal Use of Mercury.—This, as is well known, is the prevalent mode of treating syphilis in many countries. In this country (Austria) it is used only for mild cases of relapse, and we do not recommend it under any other circumstances.

℞. Hydrarg. tannic. oxidul. (oxidulated mercuric tannate), 1.50 (gr. xxiv);
Extr. laudan., 0.10 (gr. jss).—M.
Div. in dos. No. xv. D. ad capsul. nebulos.
S.: One pill after meals t. d.

℞. Hydrarg. bichlor. corros., 0.30 (gr. v);
Extr. et pulv. rad. acori, āā.
q. s. f. pil. pond., 0.20 (gr. iij).
No. xxx; consperge pulv. liquirit.
S.: One pill every evening. Increase the dose by one pill every day up to 3 or 4 pills a day.

℞. Hydrarg. iodid. flav., 1.50 (gr. xxiv);
Extr. laudan., 0.50 (gr. viij);
Pulv. et extr. gent., q. s. ft. pilul. pond., 0.20 (gr. iij).
No. l.
S.: 1 or 2 pills twice a day.

Use of the Iodids in Syphilis.

While the value of the various mercurial preparations in the treatment of syphilis is well established, and they are found to be well nigh indispensable in four-fifths of the cases, the iodids form a most useful addition to the list of specific remedies. The iodids are indicated in general glandular enlargements, in the scrofulous diathesis, in accidental nervous complications such as headache and insomnia, and in cases where a long-continued course of mercury has proved useless. We must not be deterred by the incidental effects of iodid, such as catarrh of the nose, of the conjunctivæ, and of the respiratory tract, frontal headache, and often gastric disturbances. After the treatment has been omitted for a few days and the symptoms have subsided, it may be resumed with smaller doses, until the organism gradually becomes accustomed to the drug. In the case of some patients, who at first complained of innumerable symptoms and protested vigorously against the drug, I have been able eventually to give the maximum dose, 5 to 10 to 15 g. (ʒiss to ʒiv), either by gradually increasing the dose or by varying the form.

℞. Pot. iodid., 1.5 (gr. xxiv);
Aq. destill., 100.0 (f℥iij).
S.: Divide into 2 or 3 portions and drink during the day, diluting in fresh water.

A better way is to give the patient some *potassium iodid* in a well-corked vial and let him prepare the solution himself. He can carry it about with him more easily, and the iodid, which is very sensitive to moisture, is protected from the air and from dampness. If he is to increase the dose, he has his remedy always ready at hand. Some patients prefer to add a little fruit juice to the solution in order to disguise the taste, others prepare a concentrated watery solution and take it with milk.

Sodium iodid contains less iodin than the potassium salt; it may be given in the same form; some patients take it more easily than the other preparation.

Iodoform is not so suitable for internal use on account of its disagreeable taste; it can only be taken in the form of pills.

℞. Iodof. pur., 10.0 (ʒijss);
Extr. et pulv. acor.
āā. q. s. f. pil. pond. 0.2 (ʒss).
No. c, obteg. lamin. argent. (wrap in silver foil).
S.: 2 or 3 pills morning and evening, before eating.

We have lately tried iodothyrin (Baumann), not without success (in capsules containing 1.0 (gr. xv), prepared with sacchar. lact).

Tincture of iodin, which has occasionally been given in drops in a mucilaginous decoction, is not advisable on account of the gastric disturbances which it is apt to produce.

Last in the list of iodin preparations is **ferric iodid**:

℞. Syrup. ferri iodid, 10.0 (ʒijss);
Syrup. cort. aurant.,
Syrup. simpl., āā. 20.0 (ʒv);
S.: 2 or 3 teaspoonfuls a day for adults; 1 teaspoonful for children.

It is a good remedy, either by itself or in conjunction with mercurial preparations.

Medicinal Woods.

Decoctions of sarsaparilla, lignum guaiac, etc., are to be regarded as simple cathartics, diuretics and sudorifics. Those which contain mercury or iodin, like *Decoctum Zittmanni*, are not without medicinal qualities. The latter remedy still enjoys a certain reputation. Prescribe 200 to 300 (f ℥vj to f ℥ix) *Zittmanni Decoctum fortius*, to be taken warm early in the morning. *Decoctum Zittmanni mitius* at night before retiring, usually taken cold.

If these remedies are used, the diet must be strictly regulated. Cheese, salad, fresh fruit, beer, and all inflating foods are to be avoided. The diet should consist principally of roasted meats, vegetables, tea, ham, soft-boiled eggs, and wheat bread. If the patient has more than four stools a day, the drug is too potent and the dose must be reduced. In many cases Decoctum Zittmanni is used with advantage as an adjunct to the inunction cure.

In concluding this chapter on the general treatment, let me emphasize once more that the patient must keep his room well aired and well heated, and in the mild season must spend a good deal of time in the open air. He must have plenty of good, nutritious food. My own experience leads me to condemn all starvation or dry-bread "cures," combined with sudorific baths and other remedies calculated to weaken the organism. Although an apparent cure may sometimes be effected, and the symptoms temporarily subside, the treatment is of no lasting value; I have often had occasion to observe grave relapses within a very short time after such "cures."

As **supplementary cures** we may recommend baths—sea-baths, hot baths containing iodin or sulphur, or a systematic hydrotherapy, which is so effective in regulating various morbid conditions and tends to strengthen and harden the patient.

VENEREAL ULCERS.

It is now generally conceded that there is a form of ulcer quite distinct from the initial syphilitic lesion due to infection during sexual intercourse. It is the *simple, venereal, contagious, non-indurative, soft ulcer*, the so-called *soft chancre*.

As we have already stated, this form of ulcer may be transmitted at the same time with syphilis, but it also occurs independently and has its own peculiar characters, both as regards its form and the consequences to which it gives rise.

The seats of the venereal ulcers are the external skin

and the mucous membranes of the genital region and, under certain conditions, of the rest of the body. The specific bacillus, which has been isolated during the last few years (Ducrey-Krefting, Unna-Krefting), is contained in the pus and in the tissues of the ulcer. It gives rise to other similar ulcers which appear in series in the same organism and constitute a purely local disease.

The ulcer develops almost immediately after infection has taken place through an abrasion of the skin or mucous membrane; usually within twelve to twenty hours, in exceptional cases after thirty-six to seventy-two hours. The purulent softening is at first circumscribed, but soon shows a tendency to spread, both downward and laterally, by suppurative destruction of the skin. When mixed with syphilitic products the venereal pus constitutes a virus which is doubly dangerous to the non-syphilitic organism. In such a case the venereal ulcer develops within a very short time and is followed later on by the appearance in the same situation of the initial syphilitic lesion with its characteristic infiltration and induration. In a syphilitic individual the venereal ulcer shows its usual characters; although, according to some observers, the inflammatory reaction is aggravated by the excessive irritability inherent in all syphilitic tissues.

The ulcer begins as a pustule resembling acne; in twenty-four hours it becomes filled with pus and on the following day it bursts, discharging a mass of thick pus. The ulcer now looks as if it had been cut out with a punch, it has sharply defined edges and is surrounded by a narrow zone of inflammation. If a large wound, such as results from laceration for instance, becomes infected with venereal pus, the resulting ulcer exhibits a purulent floor with sharp, inflamed, suppurating walls. The latter are frequently undermined by the action of the pus and converted into detached shreds, more or less swollen and inflamed, and loosely adherent to the margin of the ulcer. The thick pus adheres to the surrounding surface and produces new ulcers which rapidly increase in size and, by coalescing with the original ulcer and with each other, form large, irregular ulcerating areas. If the pus is carried into a sebaceous or mucous gland, deep ulcers, as large as a pea and resembling furuncula, result. The pus is discharged through the duct of the disintegrated follicle some time before the walls break down, and a deep, eroded ulcer is exposed. Ulcers which are deep seated from the beginning frequently simulate an induration before they break down and pus is discharged. If the inflamed tissues are at all hyperplastic, there is a perceptible increase in the resistance, but the sore never assumes the character of a typical syphilitic sclerosis. Induration is always a relative conception, even syphilitic ulcerations sometimes exhibit but a slight degree of induration. The rapid development, profuse suppuration, spread by infection to the surrounding parts, and the slight degree of induration are, however, sufficiently characteristic to establish the diagnosis of venereal ulcer.

The course is very simple if proper treatment is used and the pus can be removed. Within a week or two the floor of the ulcer ceases to secrete, and granulation-tissue is formed, the walls are reduced to the level of their surroundings, and cicatrization of the ulcer begins from one side or the other. If the case is carefully treated, the resulting scar is quite soft and but little depressed below the level of the skin. Sometimes, however, owing to the seat, number, and extent of the venereal ulcers and to numerous external conditions, the course as well as the termination of the disease is more serious. We have seen venereal ulcers in the foreskin, in the skin of the scrotum, groin, and surface of the opposite thigh, separated by more or less robust bridges of healthy skin, which constituted a grave condition. Under such circumstances the lymph-glands in the neighborhood are nearly always involved.

Patients whose resisting power is low, or those who are enfeebled by other disease, suffer most from this condition, as they lack the necessary energy to care for the ulcers properly from the beginning.

It has already been stated that the genitalia are the most frequent seat of venereal ulcers. *In the male* the foreskin, the inner layer near the coronary sulcus, and particularly the frenum, which is often perforated at the base and reduced to a slender thread connecting the foreskin with the head of the penis. The head itself is attacked less frequently, possibly because it contains fewer gland-ducts and is covered with smooth epidermis. *In the female* the most frequent sites are the labia majora, especially the posterior commissure, the vaginal orifice, and the remains of the hymen (fimbriæ), the urethral orifice, the perineum, anus, vagina and vaginal portion of the uterus.

In both sexes venereal ulcers are often found in the folds of the groin, on the inner aspect of the thighs, and on the symphysis. In addition we may mention the nose, mouth, tongue, nipple, navel, fingers, and other distant parts covered with hair, which may become infected by accidental contact with the purulent secretions.

If the above-mentioned characters are borne in mind—the circular shape, sharply defined edges, rapid development, and multiplicity—venereal ulcers can readily be distinguished from other processes occurring about the genitalia, such as acne, furuncula, etc. If there is no suspicion of sexual intercourse preceding the appearance of such processes, the diagnosis is quite easy. Moist papules in process of degeneration or covered with a diphtheritic secretion constitute the most frequent source of error. Again, however, the nature of the accompanying syphilitic symptoms, infiltration of neighboring lymph-glands, lesions of the skin and mucous membranes, the longer duration of syphilitic products, and, on the other hand, the suppuration in the glands which usually accompanies venereal ulcers of long standing, afford a protection against error in diagnosis (Plates 61, 62).

The **complications** which occur with venereal ulcers

are chiefly due to the rapidly spreading inflammation which accompanies the ulceration.

An ulcer situated at the edge of the foreskin often produces an inflammatory phimosis or paraphimosis. In either case gangrene may develop in the prepuce. A deep ulcer may produce a perforation in the frenum or even into the urethra.

One of the most unpleasant complications is the **inflammation of the lymphatics on the dorsum penis**; abscesses sometimes result and the skin over the abscesses is very prone to become gangrenous (*Bubonulus Nisbethii*). If there are several abscesses, the accumulated venereal pus under the skin may produce troublesome and tedious complications requiring the most energetic treatment (Plate 64).

In the female we frequently observe ulcers in the shape of fissures between the fimbriæ, deep ulcers in the ducts of Bartholin's glands, and destructive ulcerations about the urethral orifice, part of which is sometimes entirely lost by destruction of the upper wall. The posterior commissure is often the seat of large, deep ulcers, dangerous not only on account of their long duration, but also because they are apt to spread toward the rectum.

The frequent contamination with fecal matter and the painful stretching to which the parts about the anus are subjected tend to aggravate the condition in that situation. The pain prevents the patient from fighting the ulceration as vigorously as he should, so that it often spreads to the rectum.

One of the most important complications of the soft chancre is found in the inflammation of the **local lymph-glands—bubo or adenitis.**

Adenitis rarely develops within the first two weeks after the appearance of the chancre; usually it does not appear until the third or fourth week, or even later; in some cases the glands begin to become inflamed after the ulcers have completely healed. This is apt to be the case when the ulcers are small and run their course without being noticed; it is not possible in every case to refer the adenitis to the presence of a venereal ulcer. But the statement which has occasionally been made, that adenitis may develop without either the presence or previous existence of a venereal ulcer, through the uninjured skin, is not supported by the facts (primary buboes).

The nearest group of lymph-glands is always in danger of becoming involved, whatever may be the seat of the ulcers; but, as they are most frequently found in or about the genitalia, we shall confine ourselves in the following to the *lymph-glands of the inguinal region.*

As a rule, the glands of the same side are affected; but occasionally the glands of the opposite side have been observed to suppurate, which is explained by the anatomical arrangement of the lymphatic vessels and their communication with each other.

One or more glands at first become slightly enlarged; as the swelling and pain increase, especially if the part is freely used, a flat tumor with inflamed surface appears. If several glands are affected and the cellular tissue around them becomes swollen, an immovable mass of varying size develops, completely filling the inguinal region. Once the swelling has reached this stage there is little hope of resolution; usually the tumor, which may be more or less prominent, begins to fluctuate and, unless surgical aid is rendered at once, the skin softens and undergoes liquefaction-necrosis. Sometimes gangrene supervenes, and the abscess discharges a thin, purulent material mixed with detritus. The condition is very painful and may be attended with fever and marked disturbances of the general health; if left to itself after the contents of the abscess have been discharged, it persists for a long time until the cavity of the abscess clears up, the degenerated superficial portion is cast off, and scar-formation begins. In rare cases the gangrene spreads to such an extent that large areas of the skin and subcutaneous cellular tissue are destroyed, exposing the muscles of the inguinal region and even of the lower abdomen, as if they had been dissected for anatomical demonstration. Sometimes a multilocular adenitis develops by the accumulation of pus in one spot and the undermining of the subcutaneous tissue and suppuration of the nearest gland. The day after one abscess has been opened another appears in its immediate proximity. In the days of conservative surgery patients would present themselves for treatment with six or eight such abscesses, all communicating with each other by fistulæ and discharging thin, serous pus on pressure. They have been termed strumous buboes; they contain enlarged and inflamed glands or the remains of glands embedded in a stroma of connective tissue which becomes very robust in advanced cases. We once saw such a case of neglected adenitis which resulted in putrefaction and discharge of the glands in the iliac region, followed by general sepsis.

The diagnosis of such inflammatory tumors in the inguinal region can scarcely fail to be made if the rapid development and the presence of ulcers, or at least scars at the periphery (the genitalia, inner surfaces of the thighs, nates) are taken into consideration. The differential diagnosis from scrofulous glandular abscesses, although it has no practical value, can easily be made by the general condition and the presence or absence of scrofulous signs in the skin and in the other glands of the body. Secondary abscesses from the pelvic region and diffuse phlegmon from osteal or periosteal disease can easily be distinguished by the nature of the secretions, the duration of the process, and by careful examination into the condition of the bones and cartilages. There remain inguinal and femoral hernias, which only the most superficial examiner could mistake for adenitis. It is to be remembered, however, that a testicle which has become confined at the external abdominal ring may be so swollen and inflamed as to simulate an adenitis (see Plates 62, 63).

The adenitis which accompanies the primary syphilitic lesion has been treated in its proper place.

Treatment of Venereal Ulcers.—The contagious material is contained in the pus. The principal part of the treatment consists, therefore, in removing the pus and in preventing its being retained within the ulcer or conveyed to surrounding portions of the skin.

Cauterization, on account of its rapidity and efficiency in bringing about a thorough cure, has always stood in high repute as a treatment for venereal ulcers. Copper salts, carbolic acid, ferric chlorid, nitrate of silver, caustic lime, (quicklime) and the actual cautery itself have been employed. The actual cautery, or the thermocautery which now takes its place, is still the sovereign remedy. Of the others, sulphate of copper, in a solution of 1 part of sulphate of copper to 4 parts of water is the best, as it cauterizes the ulcer and the surrounding infiltrated tissues without danger of injury to the healthy parts. The sulphate of copper is to be applied three or four times a day for some time. The wound is treated for a quarter of an hour at a time with pledgets of cotton dipped in the solution, the pledgets being changed several times. Finally a pledget is squeezed dry and allowed to remain on the ulcer. Cauterization with the solid stick is less satisfactory and more painful. The swelling which follows with the reaction must be controlled with compresses. In one day a dry, bluish eschar is formed which gradually comes away during the next two days. A dry, anemic wound remains, after which granulation and cicatrization must be induced by other means.

Concentrated **carbolic acid** has a similar action; it is applied with a cotton pledget fastened to the end of a wooden applicator like a brush.

Silver nitrate, like copper sulphate, is better applied in a saturated solution than with the solid stick. **Vienna paste**[1] and **chlorid of zinc** cannot be used in all situations on account of the danger to the surrounding parts.

The best remedy of all is the **Paquelin thermocautery.**

[1] Caustic potash, 5; slaked lime, 6, mixed with alcohol to form a paste.

Whatever form of cauterization be used, local anesthesia should first be induced by some of the customary drugs, cocain, etc. (If the Paquelin thermocautery is used, ethyl chlorid must not be employed.)

If the seat of the ulcer is such that cauterization is not practicable (vaginal orifice, anus), it is replaced by careful applications of dusting-powder, ointments, or pastes, such as iodoform, dermatol, airol, salicylic acid, xeroform, etc. If such remedies are carefully and conscientiously applied, a few days will suffice to convert the ulcers into ordinary granulating wounds (stadium reparationis).

In cases where the skin becomes undermined, in fistulas, in perforations of the frenum of the labia, the abscesses are first carefully disinfected and then opened with the knife and converted into open, granulating wounds. If, as occasionally happens in venereal ulcer, gangrene develops, a line of demarcation soon forms, the ulcer as such is destroyed, and, after the gangrenous slough is cast off, an ordinary wound without any specific character remains.

The phagedenic ulcer is nothing but a rare subvariety of gangrene, with a tendency to rapid, purulent liquefaction. Such ulcers spread rapidly, have thick, irregular edges, and bleed very easily. The floor is covered with a dirty yellowish material; the surrounding tissue is inflamed and of a pale-red color. It is difficult to determine the cause of phagedenic ulcers; they usually occur in neglected cases, in anemic, debilitated individuals. The best treatment is the thermocautery, used under anesthesia.

Treatment of the Adenitis.—The treatment of venereal inflammation and abscess-formation in the glands varies somewhat for each individual case, hence we can give only a few general guiding principles.

Since most cases of adenitis are not due to infection by the venereal pus which contains the virus, but represent simply a form of symptomatic buboes, the effect of rest and the application of moderately cold compresses of aluminum acetate solution, supplemented by an ice-bag on the outside, should first be tried. The time-honored remedy of painting with tincture of iodin and galls occasionally produces good results; we do not, however, recommend a mere trial application of the tincture, as it only produces blisters and increases the pain. To have any effect the painting must be thorough; then a scab will be formed on the same day and the progress of the inflammation will thus be arrested. If an abscess forms in spite of such energetic treatment, immediate interference is indicated. The mildest treatment of adenitis consists in puncture, evacuation of the abscess, and irrigation of the cavity with a 1 per cent. solution of argentic nitrate, the wound being afterward dressed with iodoform or sterilized gauze and the dressing held in place with a spica bandage. The dressing is changed the next day. If the secretion is still purulent, the abscess is again irrigated. After two days the dressing is again changed, and so on as long as the indications continue.

If the abscesses are large and boggy and covered with a thin layer of skin, or if the roof has become gangrenous and the abscess bursts, the surrounding parts are first thoroughly cleansed, after which the hair is shaved, and the skin is again washed with ether and alcohol and irrigated with a lukewarm solution of sublimate. The attenuated covering is then removed, and the liquefied remains of glands and capsule, as well as the purulent inflammatory tissue between the glands, scraped out by one of the many methods known to modern surgery. This is the quickest way to effect a cure. The dressing can be left on for several days, to give the patient a chance to recover from the pain and shock of the operation.

The first-mentioned treatment for adenitis requires from ten days to three weeks to effect a cure; the second, owing to the gravity of the cases in which it is employed, from five to eight weeks.

SOLUTIONS FOR EXTERNAL USE.

℞. Hydrarg. chlor. corros., 0.5 (ʒjss);
Spirit. vin.,
Aq. dest., *āā.* 100.0 (f℥iij).
S.: Dilute in 5 times the quantity of water. To be used as a lotion, with cotton pledgets.

℞. Acid. carbol., 4.0 (ʒj);
Spirit. vin., 40.0 (fʒx);
Aq. dest., 160.0 (f℥v).
S.: External use.

℞. Sodii borat., 10 to 200.
S.: Lotion.

℞. Acid. salicyl., 2–4 to 200.
S.: As above.

℞. Cupr. sulphat., 1 to 5.
S.: For cauterization.

℞. Cupr. sulphat., 2–5 to 100.
S.: To be used in washing ulcers on the penis, prepuce, etc., for from 5 to 10 minutes, and for dressing.

Dusting-powders, to be used after the ulcers have been cleansed with the aboved-named solutions:

℞. Iodoform. pulv., 10.0 (ʒijss).
S.: Dusting-powder.

℞. Iodoform, 1 to 5.
With ether sulph.
S.: For spraying.

℞. Iodoform, 1 to 5;
Vaselin.
S.: Ointment. To be applied to the wound with gauze.

℞. Iodoform gauze, 20 per cent.
S.: For bandaging.

℞. Xeroform. pur., 10.0 (ʒijss).
S.: Dusting-powder.

℞. Dermatol. pulv., 10.0 (ʒijss).
S.: As above.

℞. Aristol.
S.: As above.

℞. Airol.
S.: As above.

℞. Hydrarg. ammoniat., 0.5 (gr. viij);
Vaselin, 15.0 (ʒjv).
S.: Ointment for dressing.

℞. Hydrarg. oxid. rubr., 0.1 (gr. jss);
Vaselin, 10.0 (ʒijss).
S.: Ointment for dressing torpid ulcers.

℞. Arg. nitr. fus.
S.: For touching proliferating ulcers after the use of iodoform and other remedies.

℞. Arg. nitr., 1 to 15.
S.: To be applied with pledgets of cotton instead of the solid stick.

℞. Bitum. phag. 5.0 (ʒjss);
Gyps. (selenite, a native sulphate of calcium), 30.0 (℥j).
S.: Powdered gypsum. For extensive gangrenous wounds and neglected ulcers.

GONORRHEA.

Gonorrhea, or blennorrhea, is usually localized in the mucous membrane of the genitalia. In the majority of cases infection takes place by direct contact; it may, however, be conveyed indirectly by polluted instruments, specula, clothing, bandages, towels, etc.

The cause of this form of venereal catarrh is the gonococcus of Neisser, which has been demonstrated in the secretions, in the affected mucous membranes, and in other tissues, glands, and submucous tissues involved in the process.

The results of numerous observations have established the fact that many other micro-organisms, whose special characters have not as yet been accurately described, exist normally and pathologically in the mucous membrane of the genitalia. In addition to the specific gonorrhea produced by the gonococcus of Neisser, there are other catarrhal affections of the genital mucous membrane in which these micro-organisms are also found.

The limited scope of this work forbids a more detailed discussion of these conditions, and for the same reason our description of the gonorrheal processes themselves is perforce very brief and synoptic.

It is a matter of clinical experience that contact of a healthy with a gonorrheal mucous membrane produces in from a few hours to three days alterations which suggest the probability, if not the certainty of the infectious nature of gonorrhea.

The **symptoms** consist of inflammation and hyperemia of the affected mucous membrane, coupled with a discharge, at first mucous, but rapidly becoming purulent. The patient himself early becomes aware of the condition on account of the swelling and subjective sense of burning. Gonorrhea is distinguished from all other forms of catarrh in the genital mucous membranes by its rapid spread and by the intense inflammation which marks the first week of its course.

Acute urethral gonorrhea in the male (urethritis blenorrhagica) begins with a burning sensation, felt most during micturition; the member is in a condition of semi-erection, the urethra itself is more or less increased in thickness and discharges at first a thin, watery pus, which later becomes thick and creamy, and in severe cases even mixed with blood.

If the patient is so careless as to neglect this condition, the gonorrheal inflammation spreads to the membranous and prostatic portions of the urethra and eventually to the neck of the bladder, although there may be at first a partial abatement of the inflammatory and subjective symptoms.

The intensity of the subjective symptoms increases with the spread and severity of the gonorrhea, the patient complains of pressure in the perineum and feels a constant desire to urinate (*tenesmus*) and, if the neck of the bladder is involved, is forced to urinate at intervals of half an hour to an hour. The inflammation itself and the pain which it produces, and which usually becomes worse at night, are sometimes very violent, so that the most phlegmatic individuals find them very distressing.

No difficulty is experienced in demonstrating the characteristic gonococci within the cells of the urethral secretion in any stage of the disease. Method: spread a minute quantity of the secretion on a cover-slip, allow it to dry, and pass it rapidly through the flame of an alcohol lamp two or three times (dry and fix), stain with carbol fuchsin, gentian-violet, methyl-blue, etc., wash in water, dry with bibulous paper, and mount in Canada balsam. The gonococci are decolorized by Gram's method. Demonstration by means of cultures will not be discussed at this point.

If the secretion is copious, the urine is turbid from admixture of pus and desquamated epithelium, which collect at the bottom of the vessel and form a dust-colored sediment. In order to diagnose a posterior urethritis, it is advisable to direct the patient to collect his morning urine in two separate portions. If the posterior part of the urethra is involved, both portions will be turbid.

After a time both secretion and desquamation of the mucous membrane diminish. Either the process disappears from parts of the tube, or the secretion everywhere diminishes in quantity; at all events there is only a scant discharge of creamy, whitish material from the urethra. In this stage of the disease the discharge as every one knows, is most abundant after a night's rest; the patient upon rising finds the opening of the urethra clogged, and, on squeezing, more or less pus is forced out (goutte militaire). The urine contains the characteristic *gonorrheal shreds* consisting of mucus, leukocytes, and epithelial cells from the urethra. At the same time there is an abatement in the subjective symptoms; the patient feels much better and his complaint is less troublesome. Excessive indulgence in the pleasures of Bacchus or Venus at this period of chronic gonorrhea may bring on an acute exacerbation which may closely simulate a fresh infection.

The gonorrheal process may penetrate the mucous membrane and attack the **submucosa**, or, more frequently, spread along the ducts which empty into the urethra and involve the urethral *glands* themselves. The urethra becomes studded with nodular outgrowths, irregularly disposed along its course, which may persist for some time, or break down and form periurethral abscesses. The inflammation in the submucosa sometimes spreads to the **corpus cavernosum**, where it produces tumors as large as, or larger than a walnut; if the inflammation does not subside, large abscesses result. The symptoms of such an acute peri-urethritis are very marked: pain, fever, swelling, and distortion (bending) of the penis and fluctuation.

Notwithstanding the great lengths of their ducts **Cowper's glands** are also invaded by the gonorrheal process. The swelling begins behind and at the side of the bulb of the urethra, and is attended with moderate pain in the perineum; with proper care on the part of the patient the swelling may disappear spontaneously. As, however, the causes which originally produced the inflammation in the glands, such as forced irrigation, bougies, dancing, riding, and driving, do not, as a rule, cease immediately, the inflammation usually continues for some time. A large, painful, fluctuating tumor soon develops in the perineum. The patient has high fever and sometimes chills. The abscess will burst of its own accord; but it is better not to wait for that event, as large quantities of gas and pus often accumulate in the cavity of the abscess and undermine the entire perineum as far as the anus. A speedy cure is affected by opening the abscess early; large abscesses take from two to three months to heal and usually leave a distortion of the urethra.

The evil effects of gonorrhea show themselves much more frequently in the **prostate gland** than in any of the glandular structures which have been mentioned. Acute gonorrheal prostatitis constitutes a very grave complication; the patients are feverish, complain of difficulty in urination and defecation (dysuria and tenesmus), and constant burning pain in the rectum. On palpation per rectum one or both lobes of the prostate are found to be swollen, hard, and very sensitive to pressure. If the process goes on to suppuration, the accumulated pus can sometimes be felt as a soft mass within the swollen parenchyma. In most cases it is discharged spontaneously through the urethra and leaves a flattened area in the prostate which can be distinctly felt. Sometimes the swelling does not suppurate, but results instead in a permanent enlargement of the prostate. This is a very frequent concomitant of chronic gonorrhea, and may last as long as the patient lives, without his ever being aware of it.

Another complication which is often overlooked by the patient is disease of the **seminal vesicles.** It may occur independently and run its course without being detected, or it may be accompanied by inflammation of the prostate and of the epididymis.

The most frequent complication of all is **gonorrheal epididymitis.** It makes its appearance in the third week of the disease. Usually it is unilateral, but it may affect both sides, either simultaneously or one after the other. The testicle, which is partially enclosed in the epididymis, becomes swollen and painful. The spermatic cord, containing the blood-vessels and the vas deferens, become as large as a finger and can be felt as high up as the inguinal canal. The condition may be very painful and attended with high fever, so that the least susceptible patients are compelled to take to their beds. With antiphlogistic treatment and complete rest the pain and fever disappear in from five to eight days, leaving a moderate swelling and more or less induration of the epididymis, which subsides in from four to five weeks. The scrotum is always involved in the inflammatory process and adheres to the swollen epididymis, but usually only at the inferior pole on the affected side. In some cases of epididymitis there is quite a marked exudation into the tunica vaginalis propria (*acute hydrocele*), the pressure from which causes great pain. The exudate is sometimes so copious that it can be seen through the superficial layers of the scrotum at the superior pole, as in chronic hydrocele.

In rare instances the inflammation extends to the investing membrane of the testicles; the parenchyma of the glands themselves becoming involved during the inflammatory stage of the disease only. A very few of the many cases we have seen resulted in contraction of the connective tissue and atrophy of the gland-substance itself.

In the female gonorrhea is more apt to involve the entire genito-urinary tract than in the male. Ignorance of the nature and significance of the disease on the part of the patients is largely responsible for this. The patients neglect to consult a physician, either because they do not realize the gravity of their condition or because they dread exposure. The primary seat is the mucous membrane of the vulva, vagina, and urethra, which becomes inflamed and discharges a viscous, purulent secretion. The patient complains of burning pain during micturition. The inflammation soon attacks the labia; intertrigo and eczema develop in the skin and aggravate the burning and itching, or even cause actual pain. If treatment is still delayed, the gonorrheal process soon produces acute endometritis, salpingitis, perimetritis, oöphoritis, and even peritonitis. In this way chronic gonorrheal conditions frequently develop, having their principal seat in the uterine cavity and extending in both directions at the least exacerbation of the process. The patients are tortured by fever and various kinds of pain, varying according to the seat of the inflammation; pains in the lower abdomen, dragging pains in the back and lumbar region, pains before menstruation, etc.

Gonorrheal *cystitis* and *pyelitis* are very rare even in protracted cases of gonorrhea, although urethritis is both common and persistent. They are certainly much less frequent than in the male, where these complications constitute the most obstinate and, in their consequences, often the most serious sequelæ.

The complications discussed so far may be regarded as the direct extensions of the gonorrheal process. Of late years, however, numerous clinical observations and discoveries in bacteriology have demonstrated the occurrence of a metastasis of gonococci; that is, an extension to remote organs by way of the blood- and lymph-channels. The most frequent of these complications is **gonorrheal rheumatism**. Usually the disease attacks but one joint (the knee); occasionally, however, several joints are involved at the same time. The cavity of the affected joint becomes filled with a serous exudate and considerably swollen; there is a moderate rise in the temperature and a good deal of pain. The patient keeps the affected joint rigidly fixed, and is usually confined to his bed for some time in consequence.

The joint-affection is sometimes combined with a *tenosynovitis*, or the latter may occur independently. The inflammation of the sheath is characterized by swelling, redness, and considerable pain in the affected tendons, the extremities involved being for the time disabled.

The two last-named processes are very painful while they last, and interfere with the use of the extremities for some time; they usually heal, however, without leaving any permanent alterations.

Diseases of *internal organs*, such as endocarditis or pleuritis, as the direct consequences of gonorrhea are extremely rare. On this account and because they so closely resemble diseases which depend on entirely different etiological factors the diagnosis is very difficult; after every other possibility has been excluded a presumptive diagnosis may be made, if a gonorrheal condition exists.

The **mucous membrane of the rectum** often becomes involved in the gonorrheal process, especially in female patients, either by the gonorrheal pus dripping down from the vagina, or by direct infection through accidental contact with the penis during sexual intercourse. When gonorrhea attacks the rectum it is even more distressing and painful than when it is confined to the genitalis. Defecation is difficult and painful. The constant burning pain is so intense that the patients are as restless and excitable as if they were suffering from a severe illness. A purulent discharge mixed with blood and the remains of the broken-down mucous membrane flows from the anus. Energetic treatment is required; the condition often lasts for weeks and may result in extensive cicatricial contractions of the rectum.

Condylomata Acuminata.

Gonorrheal warts, venereal papillomata, must be mentioned among the sequelæ of gonorrhea. Although it is not definitely known that the gonorrheal virus in itself is capable of producing these papillomatous growths, their frequent occurrence in old, and even in acute cases of gonorrhea justifies the assumption that they are chiefly due to the constant irritation of the mucous membrane and external skin by the gonorrheal secretion. At first the surface of the mucous membrane presents a scarcely perceptible roughening, resembling plush or the surface of a well-trimmed lawn. The process begins by a lengthening and swelling of the papillæ. As the condylomata develop, the capillary loops in the papillæ become longer and engorged with blood, and surrounded by a moderate round-celled infiltration. But the most conspicuous changes are observed in the Malpighian layer, the greatly proliferated epithelium of which covers over the papillæ and fills the intervals between them in thick layers. Owing to the constant maceration, the cells of the epidermis do not become horny, but undergo desquamation, so that the Malpighian layer is practically laid bare. At this stage the surface of the condylomata is moist and greasy to the touch. But the process is rarely limited to this cell-proliferation; usually groups of swollen papillæ unite to form tumors as large as a pea or a hazelnut, rising from the base of the skin. Extensive growths of this kind present the appearance of cauliflower, from the coalition of numerous papillomata; the surface is ragged and deeply furrowed. The degenerating epithelium and the whole mass of purulent material cause the patient much annoyance both on account of the offensive odor and the copious secretion. The capillary loops in the papillæ are easily injured, so that the blood oozes out of the fissures between the papillomata if the tumor is subjected to any injury.

In the male the condylomata occur most frequently on the inner surface of the foreskin, in the neck, and on the glans penis at the urethral orifice and sometimes as high up as the fossa navicularis. Large proliferations in the preputial sac may by their pressure produce inflammation and even necrosis of the foreskin. If the proliferations are very extensive, the differential diagnosis from carcinoma is often difficult; it is based on the duration of the process, the presence or absence of isolated papillomata near the periphery of the larger tumors and on the condition of the inguinal glands.

Condylomata acuminata are more frequent **in the female.** The vestibule, orifice of the urethra, vagina, and even the vaginal portion of the uterus are often covered with numerous proliferations, which by their rapid growth greatly annoy both patient and doctor. In addition the external skin of the female genitalia as far as the groin, and the perineum and region about the anus, may be the seat of such venereal papillomata; we have even seen tumors as large as the fist occupying this entire region.

TREATMENT OF GONORRHEA AND ITS COMPLICATIONS.

Acute Gonorrhea.

During the stage of profuse purulent secretion the following remedies are employed:

(*a*) **Internal.**—

℞. Mass. copaib., gtt. x.
Dent. tal. dos. ad caps. gelatin., No. l.
S.: 4–6 capsules daily.

℞. Ol. santali, gtt. x.
Dent. tal. dos. ad caps. gelatin., No. l.
S.: 5–8 capsules daily.

℞. Pulv. cubeb.,
Extr. cubeb. alcohol., *āā* 5.00 ℥jss.—M.
Ft. pil. No. l.
Consperge pulv. glycyrrhiz.
S.: 3 pills 3 times a day.

℞. Pulv. cubeb., 30.00 (℥j);
Extr. acori calami,
Extr. gentian., *āā* 1.00 (gr. xv).—M.
S.: One knife-point after eating, 3 times a day.

(*b*) **For Injection.**—During this period remedies are employed which are supposed, or known to possess disinfectant and bactericidal qualities, but no astringent effect on the mucous membranes.

℞. Ammon. sulphoichthyol., 1.5–2.00 (gr. xxiv–ʒss);
Aq. destill., 100.00 (f℥iij).
S.: To be injected 4 times a day.

℞. Argonin, 1.00–2.00 (gr. xv–ʒss);
Aq. destill., 100.00 (f℥iij).—M.
S.: To be injected 4 times a day.

℞. Acid. tartar., 2 : 100.
S.: To be injected 2 or 3 times a day; retain for 10 minutes.

℞. Potass. hypermang., 0.01 (gr. 1/7);
Aq. destill., 100.00 (f℥ij).—M.
S.: To be injected 4 times a day.

℞. Acid. boric., 2.00–4.00 (ʒss–ʒj);
Aq. destill., 100.00 (f℥ij).—M.
S.: To be injected 4 times a day.

At the end of the second, or the beginning of the third week *astringents* are employed.

We recommend:

℞. Zinci sulphat., 0.1–0.3 (gr. jss–gr. v);
Aq. destill., 100.00 (f℥ij).—M.
S.: To be injected 4 times a day.

℞. Zinci sulphocarbol., 1.00–3.0 (gr. xv–gr. xlv);
Aq. destill., 200.00 (f℥ij).—M.
S.: To be injected 4 times a day.

℞. Arg. nitr., 0.10–0.50 (gr. jss–viij);
Aq. destill., 200.00 (f℥vj).—M.
S.: To be injected 4 times a day.

℞. Zinci sozoiodol., 1.00–5.00 (gr. xv–ʒjss);
Aq. destill., 200.00 (f℥vj).—M.
S.: To be injected 4 times a day.

℞. Zinci sulphat., 0.5 (gr. viij);
Plumb. acetat., 1.00 (gr. xv);
Aq. destill., 200.00 (f℥ij).—M.
S.: To be injected 4 times a day. Shake before using.

℞. Bismuth. subnitr., 5.00 (ʒjss);
Aq. destill., 200.00 (f℥vj).—M.
S.: To be injected 4 times a day. Shake before using.

Chronic Gonorrhea.

Injections of ½ to 2 per cent. solutions of argentic nitrate, using Guyon's or Ultzmann's syringe. As high as 5 per cent. solutions of argentic nitrate are used for astringents.

In addition, the posterior portion of the urethra, in the male, is to be irrigated with the lotions given for acute gonorrhea, using a Nélaton catheter or Ultzmann's irrigation catheter.

Inflammation of the Neck of the Bladder (Cystitis Colli Vesicæ).

The patient must remain in bed; the bowels must be kept open and the diet regulated.

Internally:

℞. Folior. uvæ ursi;
Herniariæ glabræ, āā 25.00 (℥ij).
S.: Tea.

℞. Decoct. sem. lini, 200.00 (f℥vj);
Extr. laudani, 0.10 (gr. jss).
S.: Take one tablespoonful.

℞. Salol, 10.00 (℥ijss);
Div. in dos. æquales, No. x.
S.: 4 powders daily.

℞. Terebinth. lig.;
Extr. cinchonæ;
Magnes. carb., āā 5.0 (ʒjss);
Extr. et pulv. acori calami, āā q. s.;
F. pil. pond., 0.20 (gr. iij);
No. c, consperge pulv. aromat.
S.: 6–8 pills a day.

Prostatitis.

In the *acute* form lukewarm sitz-baths and applications of Arzberg's apparatus, either exclusively or alternating with suppositories:

℞. Morph. muriat., 0.10 (gr. jss);
Olei theobr. q. s. u. f. suppos., No. x.

℞. Extr. opii aquosi, 0.03 (gr. ½);
Olei theobr., 3.0 (gr. xlv);
F. suppos. div. dos. vi.
S.: Suppositories.

℞. Extr. bellad., 0.10 (gr. jss);
Olei theobromæ q. s. u. f. suppos., No. x.

If there is fluctuation, the abscess is opened from the rectum, using the proper surgical precautions.

In *chronic prostatitis* suppositories of:

℞. Potass. iod., 1.00 (gr. xv);
Iod. puri., 0.10 (gr. jss);
Extr. laudan., 0.15 (gr. ij);
Olei theobr. q. s. u. f. suppos., No. x.

or

℞. Potass. iod., 1.0–1.5 (gr. xv–gr. xxiv);
Solve in aq. dest., adde
mucilag. semin. cydon., 150.0 (℥ivss).
S.: To be taken as a clyster after 1 stool.

In addition lukewarm sitz-baths. The bowels must be moved daily. Massage of the prostate also has a good effect.

Epididymitis.

In acute epididymitis compresses of ice-cold water or ice-bag. The bowels must be regulated. The pain is sometimes so great that the exhibition of opiates or a hypodermatic injection of morphin becomes imperative.

℞. Bismuth. subnitr., 10.00 (ʒijss);
Extr. laudan., 0.10 (gr. jss).
F. pulv. div. in dos. æqual., No. x.
S.: 3 or 4 powders daily.

After the period of severe pain is over, inunctions with the following preparations, to reduce the thickening in epididymis and spermatic cord:

℞. Extr. bellad., 1.00 (gr. xv);
Ung. ciner.,
Ung. simpl., *āā* 10.00 (ʒijss).
Ft. ung.

or painting with

R. Tinct. iod.,
Tinct. gall., *āā*.

or with:

℞. Iod. puri., 0.02 (gr. $\frac{1}{3}$);
Potass. iod., 2.50 (gr. xl);
Ung. emollient., 25.00 (ʒij).
Ft. ung.

The patient should also be directed to wear a well-fitting suspensory. Some prefer *Fricke's adhesive-plaster dressing.*

Chronic Cystitis.

Irrigation of the bladder with:

℞. Acid boric., 50.0 (℥jss);
Aq. dest., 1000.0 (Oij).

℞. Potass hypermang., 1.0–3.0 (gr. xv–xlv);
Aq. dest., 1000.0 (Oij).

℞. Formalin, 1.0 (gr. xv);
Aq. dest., 1000.0 (Oij).

℞. Argent. nitr., 1.00–2.50 (gr. xv–xl);
Aq. dest., 2000.00–1000.00 (Oiv to Oij).

Balanitis and Balanoposthitis

constitute a frequent complication of urethral gonorrhea in the male. They also occur without urethral disease. The best remedy is to wash the penis with antiseptic lotions; if there is *phimosis,* the preputial sac should be irrigated with a 1–3 per cent. solution of copper sulphate or with a solution of aluminum acetate; in addition the part should be dusted with:

℞. Dermatoli,
Amyli, *āā.*
S.: Dusting-powder.

℞. Acidi salicyl.,
Zinci oxydati, 1.0 (gr. xv);
Amyli, *āā* 10.0 (ℨijss).
Ft. p.
S.: Dusting-powder.

In the female, gonorrhea is usually localized in the vaginal portions of the urethra, in the cervix, and in Bartholin's glands. The *urethral affection* is treated according to the same principles as in the male. In order to remove the gonorrheal secretion which flows from the cervix and collects in the vagina, it is well to irrigate the part with a lukewarm 2 per cent. solution of soda, and after that with:

℞. Cupr. sulph., 1.0 (gr. xv);
Aq. dest., 1000.0 (Oij).

℞. Alumin. crud., 5.0 (ℨjss);
Aq. dest., 1000.0 (Oij).

℞. Hydrarg. chlor. corr., 1.0 (gr. xv);
Aq. dest., 1000.0 (Oij)

℞. Ammon. sulphoichthyol., 10.0–20.0 (ʒijss–ʒv);
Aq. dest., 1000.0 (Oij).

To combat the *cervical affection* 1 per cent. solutions of argentic nitrate, tincture of iodin, and iodoform suppositories are used. The caustic substances are applied in solution with the uterine sound. In the intervals between the applications the vagina is packed with tampons dipped in the above-named substances, the tampons being frequently changed.

Erosions of the *labia* are to be sprinkled with iodoform powder, or a tampon of iodoform gauze is introduced into the vagina as far as the orifice.

In suppuration of the *ducts* of *Bartholin's glands* or of the glands themselves the abscess is to be opened, or the entire gland is extirpated. In catarrhal inflammation of the duct injections of about 1 per cent. argentic nitrate solution with Anel's syringe are to be recommended.

Other complications of gonorrhea are treated according to the same principles as in the male.

Hematuria.

The patient is put to bed and the bowels are regulated.

℞. Extr. hæmostatic., 2.0–3.0 (gr. xxx–xlv);
Aq. dest., 130.0 (f ℥ij);
Syr. acidi hydriodic., 20.0 (ʒv).
S.: One tablespoonful every two hours.

Gonorrheal Rheumatism.

As a rule, preparations of salicylic acid have no effect. This peculiarity of the gonorrheal form of the disease may be of diagnostic value in doubtful cases. We recommend potassium iodid in doses of 4 to 6 grs. per diem or

℞. Sodii citrat., 5.0 (gr. x–ʒj);
D. tal. dos. No. x.
S.: 4 powders daily.

To allay the pain, the affected joint must be kept absolutely quiet and treated with compresses wet with aluminum acetate (liquor Burow) or even with ice-bags. If the pain is very intense, especially at night, hypodermatic injections of morphin are the only effective remedy. After the inflammation has abated, the part may be dressed with a starched or silica bandage; the patient feels more comfortable and is better able to change his position. If, after the bandage is removed, the part is still swollen, warm baths, massage, and painting with tincture of iodin are to be recommended. In order to prevent a more or less permanent stiffness, the patient should be anesthetized and the adhesions severed soon after the inflammatory symptoms have subsided.

Condylomata Acuminata.

Wash with antiseptic lotions (lysol 1 per cent., carbol 2 per cent., copper sulphate 1–3 per cent.). Compresses of the alkaline earths or Burow's solution of aluminum acetate and dusting-powders:

℞. Plumb. acet. basic. crystallis. pulv.,
Alum. crud. pulveris, *āā* 10.0 (ʒijss);
Dermatol., 30.0 (℥j);
Talci Venet. subtil. pulver., 50.0 (℥xiij).

If many gonorrheal warts are present, they should be thoroughly cleansed and moistened with water and then dusted with

℞. Resorcin,
Amyl., *āā* 10.0 (ʒijss),

after which they are to be isolated with gauze until the next dressing, which should take place after five or six hours. If the groups are large, it is best to scrape them out thoroughly with a curet, and cauterize the base with chromic acid (25–50 per cent.) or with sesquichlorid of iron. Very large proliferations must be excised with their base, leaving a funnel-shaped wound which is sewed up after the bleeding has stopped. Removal with the thermocautery is often the simplest and most radical operative measure. The wound is afterward treated according to the recognized principles of disinfection and dressed with an absorbent bandage.

INDEX.

STANDARD

Medical and Surgical Works

PUBLISHED BY

W. B. SAUNDERS, 925 Walnut Street, Philadelphia, Pa.

For list of the latest publications, see page 31.

The works indicated thus (*) are sold by SUBSCRIPTION (*not by booksellers*), usually through travelling solicitors, but they can be obtained *direct* from the office of publication (charges of shipment prepaid) by remitting the quoted prices. Full *descriptive circulars* of such works will be sent to any address upon application.

All the other BOOKS advertised in this catalogue are commonly for sale by *booksellers* in all parts of the United States; but any book will be sent by the publisher to any address (post-paid) on receipt of the price herein given.

GENERAL INFORMATION.

One Price. One price absolutely without deviation. No discounts allowed, regardless of the number of books purchased at one time. Prices on all works have been fixed extremely low, with the view to selling them strictly net and for cash.

Orders. An order accompanied by remittance will receive prompt attention, books being sent to any address in the United States, by mail or express, all charges prepaid. We prefer to send books **by express** when possible, and if sent C. O. D. **we pay all charges** for returning the money. Small orders of three dollars or less **must invariably be accompanied by remittance.**

How to Send Money by Mail. There are four ways by which money can be sent at our risk, namely: a post-office money order, an express money order, a bank-check (draft), and in a registered letter. Money sent in any other way is at the sender's risk. Silver should not be sent through the mail.

Shipments. All books, being packed in patent metal-edged boxes, necessarily reach our patrons by mail or express in excellent condition.

Subscription Books. Books in our catalogue marked "For sale by subscription only" may be secured by ordering them through any of our authorized travelling salesmen, or direct from the Philadelphia office; they are **not** for sale by booksellers. All other books in our catalogue can be procured of any bookseller at the advertised prices, or directly from us. *We handle only our own publications,* and cannot supply second-hand books nor the publications of other houses.

Latest Editions. In every instance the latest revised edition is sent.

Bindings. In ordering, be careful to state the style of binding desired—Cloth, Sheep, or Half-Morocco.

Descriptive Circulars. A complete descriptive circular, giving table of contents, etc. of any book sold by subscription only, will be sent free on application.

For Sale by Subscription.

AN AMERICAN TEXT-BOOK OF PHYSIOLOGY. Edited by WILLIAM H. HOWELL, PH. D., M. D., Professor of Physiology in the Johns Hopkins University, Baltimore, Md. One handsome octavo volume of 1052 pages, fully illustrated. Prices: Cloth, $6.00 net; Sheep or Half-Morocco, $7.00 net.

This work is the most notable attempt yet made in America to combine in one volume the entire subject of Human Physiology by well-known teachers who have given especial study to that part of the subject upon which they write. The completed work represents the present status of the science of Physiology, particularly from the standpoint of the student of medicine and of the medical practitioner.

The collaboration of several teachers in the preparation of an elementary text-book of physiology is unusual, the almost invariable rule heretofore having been for a single author to write the entire book. One of the advantages to be derived from this collaboration method is that the more limited literature necessary for consultation by each author has enabled him to base his elementary account upon a comprehensive knowledge of the subject assigned to him; another, and perhaps the most important, advantage is that the student gains the point of view of a number of teachers. In a measure he reaps the same benefit as would be obtained by following courses of instruction under different teachers. The different standpoints assumed, and the differences in emphasis laid upon the various lines of procedure, chemical, physical, and anatomical, should give the student a better insight into the methods of the science as it exists to-day. The work will also be found useful to many medical practitioners who may wish to keep in touch with the development of modern physiology.

The main divisions of the subject-matter are as follows: General Physiology of Muscle and Nerve — Secretion — Chemistry of Digestion and Nutrition — Movements of the Alimentary Canal, Bladder, and Ureter—Blood and Lymph —Circulation—Respiration—Animal Heat—Central Nervous System—Special Senses — Special Muscular Mechanisms — Reproduction — Chemistry of the Animal Body.

CONTRIBUTORS:

HENRY P. BOWDITCH, M. D.,
Professor of Physiology, Harvard Medical School.

JOHN G. CURTIS, M. D.,
Professor of Physiology, Columbia University, N. Y. (College of Physicians and Surgeons).

HENRY H. DONALDSON, Ph. D.,
Head-Professor of Neurology, University of Chicago.

W. H. HOWELL, Ph. D., M. D.,
Professor of Physiology, Johns Hopkins University.

FREDERIC S. LEE, Ph. D.,
Adjunct Professor of Physiology, Columbia University, N. Y. (College of Physicians and Surgeons).

WARREN P. LOMBARD, M. D.,
Professor of Physiology, University of Michigan.

GRAHAM LUSK, Ph. D.,
Professor of Physiology, Yale Medical School.

W. T. PORTER, M. D.,
Assistant Professor of Physiology, Harvard Medical School.

EDWARD T. REICHERT, M. D.,
Professor of Physiology, University of Pennsylvania.

HENRY SEWALL, Ph. D., M. D.,
Professor of Physiology, Medical Department, University of Denver

For Sale by Subscription.

AN AMERICAN TEXT-BOOK OF APPLIED THERAPEUTICS. For the Use of Practitioners and Students. Edited by JAMES C. WILSON, M. D., Professor of the Practice of Medicine and of Clinical Medicine in the Jefferson Medical College. One handsome octavo volume of 1326 pages. Illustrated. Prices: Cloth, $7.00 net; Sheep or Half-Morocco, $8.00 net.

The arrangement of this volume has been based, so far as possible, upon modern pathologic doctrines, beginning with the intoxications, and following with infections, diseases due to internal parasites, diseases of undetermined origin, and finally the disorders of the several bodily systems—digestive, respiratory, circulatory, renal, nervous, and cutaneous. It was thought proper to include also a consideration of the disorders of pregnancy.

The list of contributors comprises the names of many who have acquired distinction as practitioners and teachers of practice, of clinical medicine, and of the specialties.

CONTRIBUTORS:

Dr. I. E. Atkinson, Baltimore, Md.
Sanger Brown, Chicago, Ill.
John B. Chapin, Philadelphia, Pa.
William C. Dabney, Charlottesville, Va.
John Chalmers DaCosta, Philada., Pa.
I. N. Danforth, Chicago, Ill.
John L. Dawson, Jr., Charleston, S. C.
F. X. Dercum, Philadelphia, Pa.
George Dock, Ann Arbor, Mich.
Robert T. Edes, Jamaica Plain, Mass.
Augustus A. Eshner, Philadelphia, Pa.
J. T. Eskridge, Denver, Col.
F. Forchheimer, Cincinnati, O.
Carl Frese, Philadelphia, Pa.
Edwin E. Graham, Philadelphia, Pa.
John Guitéras, Philadelphia, Pa.
Frederick P. Henry, Philadelphia, Pa.
Guy Hinsdale, Philadelphia, Pa.
Orville Horwitz, Philadelphia, Pa.
W. W. Johnston, Washington, D. C.
Ernest Laplace, Philadelphia, Pa.
A. Laveran, Paris, France.

Dr. James Hendrie Lloyd, Philadelphia, Pa.
John Noland Mackenzie, Baltimore, Md.
J. W. McLaughlin, Austin, Texas.
A. Lawrence Mason, Boston, Mass.
Charles K. Mills, Philadelphia, Pa.
John K. Mitchell, Philadelphia, Pa.
W. P. Northrup, New York City.
William Osler, Baltimore, Md.
Frederick A. Packard, Philadelphia, Pa.
Theophilus Parvin, Philadelphia, Pa.
Beaven Rake, London, England.
E. O. Shakespeare, Philadelphia, Pa.
Wharton Sinkler, Philadelphia, Pa.
Louis Starr, Philadelphia, Pa.
Henry W. Stelwagon, Philadelphia, Pa.
James Stewart, Montreal, Canada.
Charles G. Stockton, Buffalo, N. Y.
James Tyson, Philadelphia, Pa.
Victor C. Vaughan, Ann Arbor, Mich.
James T. Whittaker, Cincinnati, O.
J. C. Wilson, Philadelphia, Pa.

The articles, with two exceptions, are the contributions of American writers. Written from the standpoint of the practitioner, the aim of the work is to facilitate the application of knowledge to the prevention, the cure, and the alleviation of disease. The endeavor throughout has been to conform to the title of the book—Applied Therapeutics—to indicate the course of treatment to be pursued at the bedside, rather than to name a list of drugs that have been used at one time or another.

While the scientific superiority and the practical desirability of the metric system of weights and measures is admitted, it has not been deemed best to discard entirely the older system of figures, so that both sets have been given where occasion demanded.

For Sale by Subscription.

AN AMERICAN TEXT-BOOK OF OBSTETRICS. Edited by RICHARD C. NORRIS, M. D.; Art Editor, ROBERT L. DICKINSON, M. D. One handsome octavo volume of over 1000 pages, with nearly 900 colored and half-tone illustrations. Prices: Cloth, $7.00; Sheep or Half-Morocco, $8.00.

The advent of each successive volume of the *series* of the AMERICAN TEXT-BOOKS has been signalized by the most flattering comment from both the Press and the Profession. The high consideration received by these text-books, and their attainment to an authoritative position in current medical literature, have been matters of deep *international* interest, which finds its fullest expression in the demand for these publications from all parts of the civilized world.

In the preparation of the "AMERICAN TEXT-BOOK OF OBSTETRICS" the editor has called to his aid proficient collaborators whose professional prominence entitles them to recognition, and whose disquisitions exemplify **Practical Obstetrics.** While these writers were each assigned special themes for discussion, the correlation of the subject-matter is, nevertheless, such as ensures logical connection in treatment, the deductions of which thoroughly represent the latest advances in the science, and which elucidate *the best modern methods of procedure.*

The more conspicuous feature of the treatise is its wealth of illustrative matter. The production of the illustrations had been in progress for several years, under the personal supervision of Robert L. Dickinson, M. D., to whose artistic judgment and professional experience is due the **most sumptuously illustrated work of the period.** By means of the photographic art, combined with the skill of the artist and draughtsman, conventional illustration is superseded by rational methods of delineation.

Furthermore, the volume is a revelation as to the possibilities that may be reached in mechanical execution, through the unsparing hand of its publisher.

CONTRIBUTORS:

Dr. James C. Cameron.
Edward P. Davis.
Robert L. Dickinson.
Charles Warrington Earle.
James H. Etheridge.
Henry J. Garrigues.
Barton Cooke Hirst.
Charles Jewett.

Dr. Howard A. Kelly.
Richard C. Norris.
Chauncey D. Palmer.
Theophilus Parvin.
George A. Piersol.
Edward Reynolds.
Henry Schwarz.

"At first glance we are overwhelmed by the magnitude of this work in several respects, viz.: First, by the size of the volume, then by the array of eminent teachers in this department who have taken part in its production, then by the profuseness and character of the illustrations, and last, but not least, the conciseness and clearness with which the text is rendered. This is an entirely new composition, embodying the highest knowledge of the art as it stands to-day by authors who occupy the front rank in their specialty, and there are many of them. We cannot turn over these pages without being struck by the superb illustrations which adorn so many of them. We are confident that this most practical work will find instant appreciation by practitioners as well as students."—*New York Medical Times.*

Permit me to say that your American Text-Book of Obstetrics is the most magnificent medical work that I have ever seen. I congratulate you and thank you for this superb work, which alone is sufficient to place you first in the ranks of medical publishers.

With profound respect I am sincerely yours, ALEX. J. C. SKENE.

For Sale by Subscription.

AN AMERICAN TEXT-BOOK ON THE THEORY AND PRACTICE OF MEDICINE. By American Teachers. Edited by WILLIAM PEPPER, M. D., LL.D., Provost and Professor of the Theory and Practice of Medicine and of Clinical Medicine in the University of Pennsylvania. Complete in two handsome royal-octavo volumes of about 1000 pages each, with illustrations to elucidate the text wherever necessary. Price per Volume: Cloth, $5.00 net; Sheep or Half-Morocco, $6.00 net.

VOLUME I. CONTAINS:

Hygiene.—Fevers (Ephemeral, Simple Continued, Typhus, Typhoid, Epidemic Cerebrospinal Meningitis, and Relapsing).—Scarlatina, Measles, Rötheln, Variola, Varioloid, Vaccinia, Varicella, Mumps, Whooping-cough, Anthrax, Hydrophobia, Trichinosis, Actinomycosis, Glanders, and Tetanus.—Tuberculosis, Scrofula, Syphilis, Diphtheria, Erysipelas, Malaria, Cholera, and Yellow Fever.—Nervous, Muscular, and Mental Diseases etc.

VOLUME II. CONTAINS:

Urine (Chemistry and Microscopy).—Kidney and Lungs.—Air-passages (Larynx and Bronchi) and Pleura.—Pharynx, Œsophagus, Stomach and Intestines (including Intestinal Parasites), Heart, Aorta, Arteries and Veins.—Peritoneum, Liver, and Pancreas.—Diathetic Diseases (Rheumatism, Rheumatoid Arthritis, Gout, Lithæmia, and Diabetes.)—Blood and Spleen.—Inflammation, Embolism, Thrombosis, Fever, and Bacteriology.

The articles are not written as though addressed to students in lectures, but are exhaustive descriptions of diseases, with the newest facts as regards Causation, Symptomatology, Diagnosis, Prognosis, and Treatment, including a large number of approved formulæ. The recent advances made in the study of the bacterial origin of various diseases are fully described, as well as the bearing of the knowledge so gained upon prevention and cure. The subjects of Bacteriology as a whole and of Immunity are fully considered in a separate section.

Methods of diagnosis are given the most minute and careful attention, thus enabling the reader to learn the very latest methods of investigation without consulting works specially devoted to the subject.

CONTRIBUTORS:

Dr. J. S. Billings, Philadelphia.
Francis Delafield, New York.
Reginald H. Fitz, Boston.
James W. Holland, Philadelphia.
Henry M. Lyman, Chicago.
William Osler, Baltimore.
Dr. William Pepper, Philadelphia.
W. Gilman Thompson, New York.
W. H. Welch, Baltimore.
James T. Whittaker, Cincinnati.
James C. Wilson, Philadelphia.
Horatio C. Wood, Philadelphia.

"We reviewed the first volume of this work, and said: 'It is undoubtedly one of the best text-books on the practice of medicine which we possess.' A consideration of the second and last volume leads us to modify that verdict and to say that the completed work is, in our opinion, THE BEST of its kind it has ever been our fortune to see. It is complete, thorough, accurate, and clear. It is well written, well arranged, well printed, well illustrated, and well bound. It is a model of what the modern text-book should be."—*New York Medical Journal.*

"A library upon modern medical art. The work must promote the wider diffusion of sound knowledge."—*American Lancet.*

"A trusty counsellor for the practitioner or senior student, on which he may implicitly rely."—*Edinburgh Medical Journal.*

For Sale by Subscription.

AN AMERICAN TEXT-BOOK OF SURGERY. Edited by WILLIAM W. KEEN, M.D., LL.D., and J. WILLIAM WHITE, M. D., PH. D. Forming one handsome royal-octavo volume of 1250 pages (10 x 7 inches), with 500 wood-cuts in text, and 37 colored and half-tone plates, many of them engraved from original photographs and drawings furnished by the authors. Prices: Cloth, $7.00 net; Sheep or Half-Morocco, $8.00 net.

SECOND EDITION, REVISED AND ENLARGED,

With a Section devoted to "The Use of the Röntgen Rays in Surgery."

The want of a text-book which could be used by the practitioner and at the same time be recommended to the medical student has been deeply felt, especially by teachers of surgery; hence, when it was suggested to a number of these that it would be well to unite in preparing a text-book of this description, great unanimity of opinion was found to exist, and the gentlemen below named gladly consented to join in its production. While there is no distinctive American Surgery, yet America has contributed very largely to the progress of modern surgery, and among the foremost of those who have aided in developing this art and science will be found the authors of the present volume. All of them are teachers of surgery in leading medical schools and hospitals in the United States and Canada.

Especial prominence has been given to Surgical Bacteriology, a feature which is believed to be unique in a surgical text-book in the English language. Asepsis and Antisepsis have received particular attention. The text is brought well up to date in such important branches as cerebral, spinal, intestinal, and pelvic surgery, the most important and newest operations in these departments being described and illustrated.

The text of the entire book has been submitted to all the authors for their mutual criticism and revision—an idea in book-making that is entirely new and original. The book as a whole, therefore, expresses on all the important surgical topics of the day the consensus of opinion of the eminent surgeons who have joined in its preparation.

One of the most attractive features of the book is its illustrations. Very many of them are original and faithful reproductions of photographs taken directly from patients or from specimens.

CONTRIBUTORS:

Dr. Charles H. Burnett, Philadelphia.
Phineas S. Conner, Cincinnati.
Frederic S. Dennis, New York.
William W. Keen, Philadelphia.
Charles B. Nancrede, Ann Arbor, Mich.
Roswell Park, Buffalo, N. Y.
Lewis S. Pilcher, New York.

Dr. Nicholas Senn, Chicago.
Francis J. Shepherd, Montreal, Canada.
Lewis A. Stimson, New York.
William Thomson, Philadelphia.
J. Collins Warren, Boston.
J. William White, Philadelphia.

"If this text-book is a fair reflex of the present position of American surgery, we must admit it is of a very high order of merit, and that English surgeons will have to look very carefully to their laurels if they are to preserve a position in the van of surgical practice."—*London Lancet.*

For Sale by Subscription.

AN AMERICAN TEXT-BOOK OF GYNECOLOGY, MEDICAL AND SURGICAL, for the use of Students and Practitioners. Edited by J. M. BALDY, M. D. Forming a handsome royal-octavo volume, with 360 illustrations in text and 37 colored and half-tone plates. Prices: Cloth, $6.00 net; Sheep or Half-Morocco, $7.00 net.

In this volume all anatomical descriptions, excepting those essential to a clear understanding of the text, have been omitted, the illustrations being largely depended upon to elucidate the anatomy of the parts. This work, which is thoroughly practical in its teachings, is intended, as its title implies, to be a working text-book for physicians and students. A clear line of treatment has been laid down in every case, and although no attempt has been made to discuss mooted points, still the most important of these have been noted and explained. The operations recommended are fully illustrated, so that the reader, having a picture of the procedure described in the text under his eye, cannot fail to grasp the idea. All extraneous matter and discussions have been carefully excluded, the attempt being made to allow no unnecessary details to cumber the text. The subject-matter is brought up to date at every point, and the work is as nearly as possible the combined opinions of the ten specialists who figure as the authors.

The work is well illustrated throughout with wood-cuts, half-tone and colored plates, mostly selected from the authors' private collections.

CONTRIBUTORS:

Dr. Henry T. Byford.	Dr. Howard A. Kelly.
John M. Baldy.	Florian Krug.
Edwin Cragin.	E. E. Montgomery.
J. H. Etheridge.	William R. Pryor.
William Goodell.	George M. Tuttle.

"The most notable contribution to gynecological literature since 1887, ... and the most complete exponent of gynecology which we have. No subject seems to have been neglected, ... and the gynecologist and surgeon, and the general practitioner who has any desire to practise diseases of women, will find it of practical value. In the matter of illustrations and plates the book surpasses anything we have seen."—*Boston Medical and Surgical Journal.*

"A valuable addition to the literature of Gynecology. The writers are progressive, aggressive, and earnest in their convictions."—*Medical News*, Philadelphia.

"A thoroughly modern text-book, and gives reliable and well-tempered advice and instruction."—*Edinburgh Medical Journal.*

"The harmony of its conclusions and the homogeneity of its style give it an individuality which suggests a single rather than a multiple authorship."—*Annals of Surgery.*

"It must command attention and respect as a worthy representation of our advanced clinical teaching."—*American Journal of Medical Sciences.*

For Sale by Subscription.

AN AMERICAN TEXT-BOOK OF THE DISEASES OF CHILDREN. By American Teachers. Edited by LOUIS STARR, M. D., assisted by THOMPSON S. WESTCOTT, M. D. In one handsome royal-8vo volume of 1190 pages, profusely illustrated with wood-cuts, half-tone and colored plates. Net Prices: Cloth, $7.00; Sheep or Half-Morocco, $8.00.

The plan of this work embraces a series of original articles written by some sixty well-known pædiatrists, representing collectively the teachings of the most prominent medical schools and colleges of America. The work is intended to be a PRACTICAL book, suitable for constant and handy reference by the practitioner and the advanced student.

One decided innovation is the large number of authors, nearly every article being contributed by a specialist in the line on which he writes. This, while entailing considerable labor upon the editors, has resulted in the publication of a work THOROUGHLY NEW AND ABREAST OF THE TIMES.

Especial attention has been given to the latest accepted teachings upon the etiology, symptoms, pathology, diagnosis, and treatment of the disorders of children, with the introduction of many special formulæ and therapeutic procedures.

Special chapters embrace at unusual length the Diseases of the Eye, Ear, Nose and Throat, and the Skin; while the introductory chapters cover fully the important subjects of Diet, Hygiene, Exercise, Bathing, and the Chemistry of Food. Tracheotomy, Intubation, Circumcision, and such minor surgical procedures coming within the province of the medical practitioner are carefully considered.

CONTRIBUTORS:

Dr. S. S. Adams, Washington.
John Ashhurst, Jr., Philadelphia.
A. D. Blackader, Montreal, Canada.
Dillon Brown, New York.
Edward M. Buckingham, Boston.
Charles W. Burr, Philadelphia.
W. E. Casselberry, Chicago.
Henry Dwight Chapin, New York.
W. S. Christopher, Chicago.
Archibald Church, Chicago.
Floyd M. Crandall, New York.
Andrew F. Currier, New York.
Roland G. Curtin, Philadelphia
J. M. DaCosta, Philadelphia.
I. N. Danforth, Chicago.
Edward P. Davis, Philadelphia.
John B. Deaver, Philadelphia.
G. E. de Schweinitz, Philadelphia.
John Dorning, New York.
Charles Warrington Earle, Chicago.
Wm. A. Edwards, San Diego, Cal.
F. Forchheimer, Cincinnati.
J. Henry Fruitnight, New York.
Landon Carter Gray, New York.
J. P. Crozer Griffith, Philadelphia.
W. A. Hardaway, St. Louis.
M. P Hatfield, Chicago.
Barton Cooke Hirst, Philadelphia.
H. Illoway, Cincinnati.
Henry Jackson, Boston.
Charles G. Jennings, Detroit.
Henry Koplik, New York.

Dr. Thomas S. Latimer, Baltimore.
Albert R. Leeds, Hoboken, N. J.
J. Hendrie Lloyd, Philadelphia.
George Roe Lockwood, New York.
Henry M. Lyman, Chicago.
Francis T. Miles, Baltimore.
Charles K. Mills, Philadelphia.
John H. Musser, Philadelphia.
Thomas R. Neilson, Philadelphia.
W. P. Northrup, New York.
William Osler, Baltimore.
Frederick A. Packard, Philadelphia.
William Pepper, Philadelphia.
Frederick Peterson, New York.
W. T. Plant, Syracuse, New York.
William M. Powell, Atlantic City.
B. Alexander Randall, Philadelphia.
Edward O. Shakespeare, Philadelphia
F. C. Shattuck, Boston.
J. Lewis Smith, New York.
Louis Starr, Philadelphia.
M. Allen Starr, New York.
J. Madison Taylor, Philadelphia.
Charles W. Townsend, Boston.
James Tyson, Philadelphia.
W. S. Thayer, Baltimore.
Victor C. Vaughan, Ann Arbor, Mich
Thompson S. Westcott, Philadelphia.
Henry R. Wharton, Philadelphia.
J. William White, Philadelphia.
J. C. Wilson, Philadelphia.

A NEW PRONOUNCING DICTIONARY OF MEDICINE, with Phonetic Pronunciation, Accentuation, Etymology, etc. By JOHN M. KEATING, M. D., LL.D., Fellow of the College of Physicians of Philadelphia; Vice-President of the American Pædiatric Society; Ex-President of the Association of Life Insurance Medical Directors; Editor "Cyclopædia of the Diseases of Children," etc.; and HENRY HAMILTON, author of "A New Translation of Virgil's Æneid into English Rhyme;" co-author of "Saunders' Medical Lexicon," etc.; with the Collaboration of J. CHALMERS DACOSTA, M. D., and FREDERICK A. PACKARD, M. D. With an Appendix containing important Tables of Bacilli, Micrococci, Leucomaïnes, Ptomaïnes, Drugs and Materials used in Antiseptic Surgery, Poisons and their Antidotes, Weights and Measures, Thermometric Scales, New Official and Unofficial Drugs, etc. One very attractive volume of over 800 pages. Second Revised Edition. Prices: Cloth, $5.00 net; Sheep or Half-Morocco, $6.00 net; with Denison's Patent Ready-Reference Index; without patent index, Cloth, $4.00 net; Sheep or Half-Morocco, $5.00 net.

PROFESSIONAL OPINIONS.

"I am much pleased with Keating's Dictionary, and shall take pleasure in recommending it to my classes."

HENRY M. LYMAN, M. D.,
Professor of Principles and Practice of Medicine, Rush Medical College, Chicago, Ill.

"I am convinced that it will be a very valuable adjunct to my study-table, convenient in size and sufficiently full for ordinary use."

C. A. LINDSLEY, M. D.,
Professor of Theory and Practice of Medicine, Medical Dept. Yale University; Secretary Connecticut State Board of Health, New Haven, Conn.

AUTOBIOGRAPHY OF SAMUEL D. GROSS, M. D., Emeritus Professor of Surgery in the Jefferson Medical College of Philadelphia, with Reminiscences of His Times and Contemporaries. Edited by his sons, SAMUEL W. GROSS, M. D., LL.D., late Professor of Principles of Surgery and of Clinical Surgery in the Jefferson Medical College, and A. HALLER GROSS, A. M., of the Philadelphia Bar. Preceded by a Memoir of Dr. Gross, by the late Austin Flint, M. D., LL.D. In two handsome volumes, each containing over 400 pages, demy 8vo, extra cloth, gilt tops, with fine Frontispiece engraved on steel. Price per Volume, $2.50 net.

This autobiography, which was continued by the late eminent surgeon until within three months of his death, contains a full and accurate history of his early struggles, trials, and subsequent successes, told in a singularly interesting and charming manner, and embraces short and graphic pen-portraits of many of the most distinguished men—surgeons, physicians, divines, lawyers, statesmen, scientists, etc.—with whom he was brought in contact in America and in Europe; the whole forming a retrospect of more than three-quarters of a century.

SURGICAL PATHOLOGY AND THERAPEUTICS. By JOHN COLLINS WARREN, M. D., LL.D., Professor of Surgery, Medical Department Harvard University; Surgeon to the Massachusetts General Hospital, etc. A handsome octavo volume of 832 pages, with 136 relief and lithographic illustrations, 33 of which are printed in colors, and all of which were drawn by William J. Kaula from original specimens. Prices: Cloth, $6.00 net; Half-Morocco, $7.00 net.

"The volume is for the bedside, the amphitheatre, and the ward. It deals with things not as we see them through the microscope alone, but as the practitioner sees their effect in his patients; not only as they appear in and affect culture-media, but also as they influence the human body; and, following up the demonstrations of the nature of diseases, the author points out their logical treatment." (*New York Medical Journal*). "It is the handsomest specimen of book-making * * * that has ever been issued from the American medical press" (*American Journal of the Medical Sciences*, Philadelphia).

Without Exception, the Illustrations are the Best ever Seen in a Work of this Kind.

"A most striking and very excellent feature of this book is its illustrations. Without exception, from the point of accuracy and artistic merit, they are the best ever seen in a work of this kind. * * * Many of those representing microscopic pictures are so perfect in their coloring and detail as almost to give the beholder the impression that he is looking down the barrel of a microscope at a well-mounted section."—*Annals of Surgery*, Philadelphia.

PATHOLOGY AND SURGICAL TREATMENT OF TUMORS. By N. SENN, M. D., Ph. D., LL. D., Professor of Practice of Surgery and of Clinical Surgery, Rush Medical College; Professor of Surgery, Chicago Polyclinic; Attending Surgeon to Presbyterian Hospital; Surgeon-in-Chief, St. Joseph's Hospital, Chicago. One volume of 710 pages, with 515 engravings, including full-page colored plates. Prices: Cloth, $6.00 net; Half-Morocco, $7.00 net.

Books specially devoted to this subject are few, and in our text-books and systems of surgery this part of surgical pathology is usually condensed to a degree incompatible with its scientific and clinical importance. The author spent many years in collecting the material for this work, and has taken great pains to present it in a manner that should prove useful as a text-book for the student, a work of reference for the busy practitioner, and a reliable, safe guide for the surgeon. The more difficult operations are fully described and illustrated. More than *one hundred* of the illustrations are original, while the remainder were selected from books and medical journals not readily accessible.

"The most exhaustive of any recent book in English on this subject. It is well illustrated, and will doubtless remain as the principal monograph on the subject in our language for some years. The book is handsomely illustrated and printed, ... and the author has given a notable and lasting contribution to surgery."—*Journal of American Medical Association*, Chicago.

MEDICAL DIAGNOSIS. By Dr. OSWALD VIERORDT, Professor of Medicine at the University of Heidelberg. Translated, with additions, from the Second Enlarged German Edition, with the author's permission, by FRANCIS H. STUART, A. M., M. D. Third and Revised Edition. In one handsome royal-octavo volume of 700 pages, 178 fine wood-cuts in text, many of which are in colors. Prices: Cloth, $4.00 net; Sheep or Half-Morocco, $5.00 net.

In this work, as in no other hitherto published, are given full and accurate explanations of the phenomena observed at the bedside. It is distinctly a clinical work by a master teacher, characterized by thoroughness, fulness, and accuracy. It is a mine of information upon the points that are so often passed over without explanation. Especial attention has been given to the germ-theory as a factor in the origin of disease.

This valuable work is now published in German, English, Russian, and Italian. The issue of a third American edition within two years indicates the favor with which it has been received by the profession.

THE PICTORIAL ATLAS OF SKIN DISEASES AND SYPHILITIC AFFECTIONS. (American Edition.) Translation from the French. Edited by J. J. PRINGLE, M. B., F. R. C. P., Assistant Physician to, and Physician to the department for Diseases of the Skin at, the Middlesex Hospital, London. Photo-lithochromes from the famous models of dermatological and syphilitic cases in the Museum of the Saint-Louis Hospital, Paris, with explanatory wood-cuts and letter-press. In 12 Parts, at $3.00 per Part. Parts 1 to 8 now ready.

"The plates are beautifully executed."—JONATHAN HUTCHINSON, M. D. (London Hospital).

"The plates in this Atlas are remarkably accurate and artistic reproductions of *typical* examples of skin disease. The work will be of great value to the practitioner and student."—WILLIAM ANDERSON, M. D. (St. Thomas Hospital).

"If the succeeding parts of this Atlas are to be similar to Part 1, now before us, we have no hesitation in cordially recommending it to the favorable notice of our readers as one of the finest dermatological atlases with which we are acquainted."—*Glasgow Medical Journal*, Aug., 1895.

"Of all the atlases of skin diseases which have been published in recent years, the present one promises to be of greatest interest and value, especially from the standpoint of the general practitioner."—*American Medico-Surgical Bulletin*, Feb. 22, 1896.

"The introduction of explanatory wood-cuts in the text is a novel and most important feature which greatly furthers the easier understanding of the excellent plates, than which nothing, we venture to say, has been seen better in point of correctness, beauty, and general merit."—*New York Medical Journal*, Feb. 15, 1896.

"An interesting feature of the Atlas is the descriptive text, which is written for each picture by the physician who treated the case or at whose instigation the models have been made. We predict for this truly beautiful work a large circulation in all parts of the medical world where the names *St. Louis* and *Baretta* have preceded it."—*Medical Record*, N. Y., Feb. 1, 1896.

PRACTICAL POINTS IN NURSING. For Nurses in Private Practice. By EMILY A. M. STONEY, Graduate of the Training-School for Nurses, Lawrence, Mass.; Superintendent of the Training-School for Nurses, Carney Hospital, South Boston, Mass. 456 pages, handsomely illustrated with 73 engravings in the text, and 9 colored and half-tone plates. Cloth. Price, $1.75 net.

SECOND EDITION, THOROUGHLY REVISED.

In this volume the author explains, in popular language and in the shortest possible form, the entire range of *private* nursing as distinguished from *hospital* nursing, and the nurse is instructed how best to meet the various emergencies of medical and surgical cases when distant from medical or surgical aid or when thrown on her own resources.

An especially valuable feature of the work will be found in the directions to the nurse how to *improvise* everything ordinarily needed in the sick-room, where the embarrassment of the nurse, owing to the want of proper appliances, is frequently extreme.

The work has been logically divided into the following sections:

I. The Nurse: her responsibilities, qualifications, equipment, etc.
II. The Sick-Room: its selection, preparation, and management.
III. The Patient: duties of the nurse in medical, surgical, obstetric, and gynecologic cases.
IV. Nursing in Accidents and Emergencies.
V. Nursing in Special Medical Cases.
VI. Nursing of the New-born and Sick Children.
VII. Physiology and Descriptive Anatomy.

The APPENDIX contains much information in compact form that will be found of great value to the nurse, including Rules for Feeding the Sick; Recipes for Invalid Foods and Beverages; Tables of Weights and Measures; Table for Computing the Date of Labor; List of Abbreviations; Dose-List; and a full and complete Glossary of Medical Terms and Nursing Treatment.

"This is a well-written, eminently practical volume, which covers the entire range of private nursing as distinguished from hospital nursing, and instructs the nurse how best to meet the various emergencies which may arise and how to prepare everything ordinarily needed in the illness of her patient."—*American Journal of Obstetrics and Diseases of Women and Children*, Aug., 1896.

A TEXT-BOOK OF BACTERIOLOGY, including the Etiology and Prevention of Infective Diseases and an account of Yeasts and Moulds, Hæmatozoa, and Psorosperms. By EDGAR M. CROOKSHANK, M. B., Professor of Comparative Pathology and Bacteriology, King's College, London. A handsome octavo volume of 700 pages, with 273 engravings in the text, and 22 original and colored plates. Price, $6.50 net.

This book, though nominally a Fourth Edition of Professor Crookshank's "MANUAL OF BACTERIOLOGY," is practically a new work, the old one having been reconstructed, greatly enlarged, revised throughout, and largely rewritten, forming a text-book for the Bacteriological Laboratory, for Medical Officers of Health, and for Veterinary Inspectors.

A TEXT-BOOK OF HISTOLOGY, DESCRIPTIVE AND PRACTICAL. For the Use of Students. By ARTHUR CLARKSON, M. B., C. M., Edin., formerly Demonstrator of Physiology in the Owen's College, Manchester; late Demonstrator of Physiology in the Yorkshire College, Leeds. Large 8vo, 554 pages, with 22 engravings in the text, and 174 beautifully colored original illustrations. Price, strongly bound in Cloth, $6.00 net.

The purpose of the writer in this work has been to furnish the student of Histology, in one volume, with both the descriptive and the practical part of the science. The first two chapters are devoted to the consideration of the general methods of Histology; subsequently, in each chapter, the structure of the tissue or organ is first systematically described, the student is then taken tutorially over the specimens illustrating it, and, finally, an appendix affords a short note of the methods of preparation.

"We would most cordially recommend it to all students of histology."—*Dublin Medical Journal.*

"It is pleasant to give unqualified praise to the colored illustrations; . . . the standard is high, and many of them are not only extremely beautiful, but very clear and demonstrative. . . . The plan of the book is excellent."—*Liverpool Medical Journal.*

ARCHIVES OF CLINICAL SKIAGRAPHY. By SYDNEY ROWLAND, B. A., Camb. A series of collotype illustrations, with descriptive text, illustrating the applications of the New Photography to Medicine and Surgery. Price, per Part, $1.00. Parts I. to V. now ready.

The object of this publication is to put on record in permanent form some of the most striking applications of the new photography to the needs of Medicine and Surgery.

The progress of this new art has been so rapid that, although Prof. Röntgen's discovery is only a thing of yesterday, it has already taken its place among the approved and accepted aids to diagnosis.

WATER AND WATER SUPPLIES. By JOHN C. THRESH, D. SC., M. B., D. P. H., Lecturer on Public Health, King's College, London; Editor of the "Journal of State Medicine," etc. 12mo, 438 pages, illustrated. Handsomely bound in Cloth, with gold side and back stamps. Price, $2.25 net.

This work will furnish any one interested in public health the information requisite for forming an opinion as to whether any supply or proposed supply is sufficiently wholesome and abundant, and whether the cost can be considered reasonable.

The work does not pretend to be a treatise on Engineering, yet it contains sufficient detail to enable any one who has studied it to consider intelligently any scheme which may be submitted for supplying a community with water.

DISEASES OF THE EYE. A Hand-Book of Ophthalmic Practice. By G. E. DE SCHWEINITZ, M. D., Professor of Ophthalmology in the Jefferson Medical College, Philadelphia, etc. A handsome royal-octavo volume of 679 pages, with 256 fine illustrations, many of which are original, and 2 chromo-lithographic plates. Prices: Cloth, $4.00 net; Sheep or Half-Morocco, $5.00 net.

The object of this work is to present to the student, and to the practitioner who is beginning work in the fields of ophthalmology, a plain description of the optical defects and diseases of the eye. To this end special attention has been paid to the clinical side of the question; and the method of examination, the symptomatology leading to a diagnosis, and the treatment of the various ocular defects have been brought into prominence.

SECOND EDITION, REVISED AND GREATLY ENLARGED.

The entire book has been thoroughly revised. In addition to this general revision, special paragraphs on the following new matter have been introduced: Filamentous Keratitis, Blood-staining of the Cornea, Essential Phthisis Bulbi, Foreign Bodies in the Lens, Circinate Retinitis, Symmetrical Changes at the Macula Lutea in Infancy, Hyaline Bodies in the Papilla, Monocular Diplopia, Subconjunctival Injections of Germicides, Infiltration-Anæsthesia, and Sterilization of Collyria. Brief mention of Ophthalmia Nodosa, Electric Ophthalmia, and Angioid Streaks in the Retina also finds place. An Appendix has been added, containing a full description of the method of determining the corneal astigmatism with the ophthalmometer of Javal and Schiötz, and the rotations of the eyes with the tropometer of Stevens. The chapter on Operations has been enlarged and rewritten.

"A clearly written, comprehensive manual. . . . One which we can commend to students as a reliable text-book, written with an evident knowledge of the wants of those entering upon the study of this special branch of medical science."—*British Medical Journal.*

"The work is characterized by a lucidity of expression which leaves the reader in no doubt as to the meaning of the language employed. . . . We know of no work in which these diseases are dealt with more satisfactorily, and indications for treatment more clearly given, and in harmony with the practice of the most advanced ophthalmologists."—*Maritime Medical News.*

"It is hardly too much to say that for the student and practitioner beginning the study of Ophthalmology, it is the best single volume at present published."—*Medical News.*

"The latest and one of the best books on Ophthalmology. The book is thoroughly up to date, and is certainly a work which not only commends itself to the student, but is a ready reference for the busy practitioner."—*International Medical Review.*

PROFESSIONAL OPINIONS.

"A work that will meet the requirements not only of the specialist, but of the general practitioner in a rare degree. I am satisfied that unusual success awaits it."

WILLIAM PEPPER, M. D.

Provost and Professor of Theory and Practice of Medicine and Clinical Medicine in the University of Pennsylvania.

"Contains in concise and reliable form the accepted views of Ophthalmic Science."

WILLIAM THOMSON, M. D.,

Professor of Ophthalmology, Jefferson Medical College, Philadelphia, Pa.

TEXT-BOOK UPON THE PATHOGENIC BACTERIA. Specially written for Students of Medicine. By JOSEPH MCFARLAND, M. D., Professor of Pathology and Bacteriology in the Medico-Chirurgical College of Philadelphia, etc. 359 pages, finely illustrated. Price, Cloth, $2.50 net.

The book presents a concise account of the technical procedures necessary in the study of Bacteriology. It describes the life-history of pathogenic bacteria, and the pathological lesions following invasion.

The work is intended to be a text-book for the medical student and for the practitioner who has had no recent laboratory training in this department of medical science. The instructions given as to needed apparatus, cultures, stainings, microscopic examinations, etc., are ample for the student's needs, and will afford to the physician much information that will interest and profit him relative to a subject which modern science shows to go far in explaining the etiology of many diseased conditions.

The illustrations have been gathered from standard sources, and comprise the best and most complete aggregation extant.

"It is excellently adapted for the medical students and practitioners for whom it is avowedly written. . . . The descriptions given are accurate and readable, and the book should prove useful to those for whom it is written.—*London Lancet*, Aug. 29, 1896.

"The author has succeded admirably in presenting the essential details of bacteriological technics, together with a judiciously chosen summary of our present knowledge of pathogenic bacteria. . . . The work, we think, should have a wide circulation among English-speaking students of medicine."—*N. Y. Medical Journal*, April 4, 1896.

"The book will be found of considerable use by medical men who have not had a special bacteriological training, and who desire to understand this important branch of medical science."—*Edinburgh Medical Journal*, July, 1896.

LABORATORY GUIDE FOR THE BACTERIOLOGIST. By LANGDON FROTHINGHAM, M. D. V., Assistant in Bacteriology and Veterinary Science, Sheffield Scientific School, Yale University. Illustrated. Price, Cloth, 75 cents.

The technical methods involved in bacteria-culture, methods of staining, and microscopical study are fully described and arranged as simply and concisely as possible. The book is especially intended for use in laboratory work

"It is a convenient and useful little work, and will more than repay the outlay necessary for its purchase in the saving of time which would otherwise be consumed in looking up the various points of technique so clearly and concisely laid down in its pages."—*American Med.-Surg. Bulletin*.

FEEDING IN EARLY INFANCY. By ARTHUR V. MEIGS, M. D. Bound in limp cloth, flush edges. Price, 25 cents net.

SYNOPSIS: Analyses of Milk—Importance of the Subject of Feeding in Early Infancy—Proportion of Casein and Sugar in Human Milk—Time to Begin Artificial Feeding of Infants—Amount of Food to be Administered at Each Feeding—Intervals between Feedings—Increase in Amount of Food at Different Periods of Infant Development—Unsuitableness of Condensed Milk as a Substitute for Mother's Milk—Objections to Sterilization or "Pasteurization" of Milk—Advances made in the Method of Artificial Feeding of Infants.

ESSENTIALS OF ANATOMY AND MANUAL OF PRACTICAL DISSECTION, containing "Hints on Dissection." By CHARLES B. NANCREDE, M. D., Professor of Surgery and Clinical Surgery in the University of Michigan, Ann Arbor; Corresponding Member of the Royal Academy of Medicine, Rome, Italy; late Surgeon Jefferson Medical College, etc. Fourth and revised edition. Post 8vo, over 500 pages, with handsome full-page lithographic plates in colors, and over 200 illustrations. Price: Extra Cloth or Oilcloth for the dissection-room, $2.00 net.

Neither pains nor expense has been spared to make this work the most exhaustive yet concise Student's Manual of Anatomy and Dissection ever published, either in America or in Europe.

The colored plates are designed to aid the student in dissecting the muscles, arteries, veins, and nerves. The wood-cuts have all been specially drawn and engraved, and an Appendix added containing 60 illustrations representing the structure of the entire human skeleton, the whole being based on the eleventh edition of Gray's *Anatomy*.

"The plates are of more than ordinary excellence, and are of especial value to students in their work in the dissecting-room."—*Journal of American Medical Association.*

"Should be in the hands of every medical student."—*Cleveland Medical Gazette.*

"A concise and judicious work."—*Buffalo Medical and Surgical Journal.*

A MANUAL OF PRACTICE OF MEDICINE. By A. A. STEVENS, A. M., M. D., Instructor of Physical Diagnosis in the University of Pennsylvania, and Demonstrator of Pathology in the Woman's Medical College of Philadelphia. Specially intended for students preparing for graduation and hospital examinations, and includes the following sections: General Diseases, Diseases of the Digestive Organs, Diseases of the Respiratory System, Diseases of the Circulatory System, Diseases of the Nervous System, Diseases of the Blood, Diseases of the Kidneys, and Diseases of the Skin. Each section is prefaced by a chapter on General Symptomatology. Post 8vo, 512 pages. Numerous illustrations and selected formulæ. Price, $2.50. Bound in flexible leather.

FOURTH EDITION, REVISED AND ENLARGED.

Contributions to the science of medicine have poured in so rapidly during the last quarter of a century that it is well-nigh impossible for the student, with the limited time at his disposal, to master elaborate treatises or to cull from them that knowledge which is absolutely essential. From an extended experience in teaching, the author has been enabled, by classification, to group allied symptoms, and by the judicious elimination of theories and redundant explanations to bring within a comparatively small compass a complete outline of the practice of medicine.

MANUAL OF MATERIA MEDICA AND THERAPEUTICS.

By A. A. STEVENS, A. M., M. D., Instructor of Physical Diagnosis in the University of Pennsylvania, and Demonstrator of Pathology in the Woman's Medical College of Philadelphia. 445 pages. Price, Cloth, $2.25.

SECOND EDITION, REVISED.

This wholly new volume, which is based on the last edition of the *Pharmacopœia*, comprehends the following sections: Physiological Action of Drugs; Drugs; Remedial Measures other than Drugs; Applied Therapeutics; Incompatibility in Prescriptions; Table of Doses; Index of Drugs; and Index of Diseases; the treatment being elucidated by more than two hundred formulæ.

"The author is to be congratulated upon having presented the medical student with as accurate a manual of therapeutics as it is possible to prepare."—*Therapeutic Gazette.*

"Far superior to most of its class; in fact, it is very good. Moreover, the book is reliable and accurate."—*New York Medical Journal.*

"The author has faithfully presented modern therapeutics in a comprehensive work, . . . and it will be found a reliable guide."—*University Medical Magazine.*

NOTES ON THE NEWER REMEDIES: their Therapeutic Applications and Modes of Administration.

By DAVID CERNA, M. D., PH. D., Demonstrator of and Lecturer on Experimental Therapeutics in the University of Pennsylvania. Post-octavo, 253 pages. Price, $1.25.

SECOND EDITION, RE-WRITTEN AND GREATLY ENLARGED.

The work takes up in alphabetical order all the newer remedies, giving their physical properties, solubility, therapeutic applications, administration, and chemical formula.

It thus forms a very valuable addition to the various works on therapeutics now in existence.

Chemists are so multiplying compounds, that, if each compound is to be thoroughly studied, investigations must be carried far enough to determine the practical importance of the new agents.

"Especially valuable because of its completeness, its accuracy, its systematic consideration of the properties and therapy of many remedies of which doctors generally know but little, expressed in a brief yet terse manner."—*Chicago Clinical Review.*

TEMPERATURE CHART.

Prepared by D. T. LAINÉ, M. D. Size 8 × 13½ inches. Price, per pad of 25 charts, 50 cents.

A conveniently arranged chart for recording Temperature, with columns for daily amounts of Urinary and Fecal Excretions, Food, Remarks, etc. On the back of each chart is given in full the method of Brand in the treatment of Typhoid Fever.

SAUNDERS' **POCKET MEDICAL LEXICON; or, Dictionary of Terms and Words used in Medicine and Surgery.** By JOHN M. KEATING, M. D., editor of "Cyclopædia of Diseases of Children," etc.; author of the "New Pronouncing Dictionary of Medicine;" and HENRY HAMILTON, author of "A New Translation of Virgil's Æneid into English Verse;" co-author of a "New Pronouncing Dictionary of Medicine." A new and revised edition. 32mo, 282 pages. Prices: Cloth, 75 cents; Leather Tucks, $1.00.

This new and comprehensive work of reference is the outcome of a demand for a more modern handbook of its class than those at present on the market, which, dating as they do from 1855 to 1884, are of but trifling use to the student by their not containing the hundreds of new words now used in current literature, especially those relating to Electricity and Bacteriology.

"Remarkably accurate in terminology, accentuation, and definition."—*Journal of American Medical Association.*

"Brief, yet complete . . . . it contains the very latest nomenclature in even the newest departments of medicine."—*New York Medical Record.*

SAUNDERS' **POCKET MEDICAL FORMULARY.** By WILLIAM M. POWELL, M. D., Attending Physician to the Mercer House for Invalid Women at Atlantic City. Containing 1750 Formulæ, selected from several hundred of the best-known authorities. Forming a handsome and convenient pocket companion of nearly 300 printed pages, with blank leaves for Additions; with an Appendix containing Posological Table, Formulæ and Doses for Hypodermatic Medication, Poisons and their Antidotes, Diameters of the Female Pelvis and Fœtal Head, Obstetrical Table, Diet List for Various Diseases, Materials and Drugs used in Antiseptic Surgery, Treatment of Asphyxia from Drowning, Surgical Remembrancer, Tables of Incompatibles, Eruptive Fevers, Weights and Measures, etc. Third edition, revised and greatly enlarged. Handsomely bound in morocco, with side index, wallet, and flap. Price, $1.75 net.

A concise, clear, and correct record of the many hundreds of famous formulæ which are found scattered through the works of the *most eminent physicians and surgeons* of the world. The work is helpful to the student and practitioner alike, as through it they become acquainted with numerous formulæ which are not found in text-books, but have been collected from among *the rising generation of the profession, college professors, and hospital physicians and surgeons.*

"This little book, that can be conveniently carried in the pocket, contains an immense amount of material. It is very useful, and as the name of the author of each prescription is given is unusually reliable."—*New York Medical Record.*

"Designed to be of immense help to the general practitioner in the exercise of his daily calling."—*Boston Medical and Surgical Journal.*

DISEASES OF WOMEN. By HENRY J. GARRIGUES, A. M., M. D., Professor of Gynecology and Obstetrics in the New York School of Clinical Medicine; Gynecologist to St. Mark's Hospital and to the German Dispensary, New York City. In one handsome octavo volume of 728 pages, illustrated by 335 engravings and colored plates. Prices: Cloth, $4.00 net; Sheep or Half Morocco, $5.00 net.

A PRACTICAL work on gynecology for the use of students and practitioners, written in a terse and concise manner. The importance of a thorough knowledge of the anatomy of the female pelvic organs has been fully recognized by the author, and considerable space has been devoted to the subject. The chapters on Operations and on Treatment are thoroughly modern, and are based upon the large hospital and private practice of the author. The text is elucidated by a large number of illustrations and colored plates, many of them being original, and forming a complete atlas for studying *embryology* and the *anatomy* of the *female genitalia*, besides exemplifying, whenever needed, morbid conditions, instruments, apparatus, and operations.

Second Edition, Thoroughly Revised.

The first edition of this work met with a most appreciative reception by the medical press and profession both in this country and abroad, and was adopted as a text-book or recommended as a book of reference by nearly *one hundred* colleges in the United States and Canada. The author has availed himself of the opportunity afforded by this revision to embody the latest approved advances in the treatment employed in this important branch of Medicine. He has also more extensively expressed his own opinion on the comparative value of the different methods of treatment employed.

"One of the best text-books for students and practitioners which has been published in the English language; it is condensed, clear, and comprehensive. The profound learning and great clinical experience of the distinguished author find expression in this book in a most attractive and instructive form. Young practitioners, to whom experienced consultants may not be available, will find in this book invaluable counsel and help."

THAD. A. REAMY, M. D., LL.D.,
Professor of Clinical Gynecology, Medical College of Ohio; Gynecologist to the Good Samaritan and Cincinnati Hospitals.

A SYLLABUS OF GYNECOLOGY, arranged in conformity with **"An American Text-Book of Gynecology."** By J. W. LONG, M. D., Professor of Diseases of Women and Children, Medical College of Virginia, etc. Price, Cloth (interleaved), $1.00 net.

Based upon the teaching and methods laid down in the larger work, this will not only be useful as a supplementary volume, but to those who do not already possess the text-book it will also have an independent value as an aid to the practitioner in gynecological work, and to the student as a guide in the lecture-room, as the subject is presented in a manner at once systematic, clear, succinct, and practical.

A MANUAL OF PHYSIOLOGY, with Practical Exercises. For Students and Practitioners. By G. N. STEWART, M. A., M. D., D. Sc., lately Examiner in Physiology, University of Aberdeen, and of the New Museums, Cambridge University; Professor of Physiology in the Western Reserve University, Cleveland, Ohio. Handsome octavo volume of 800 pages, with 278 illustrations in the text, and 5 colored plates. Price, Cloth, $3.50 net.

"It will make its way by sheer force of merit, and *amply deserves to do so. It is one of the very best English text-books on the subject.*"—*London Lancet.*

"Of the many text-books of physiology published, we do not know of one that so nearly comes up to the ideal as does Professor Stewart's volume."—*British Medical Journal.*

ESSENTIALS OF PHYSICAL DIAGNOSIS OF THE THORAX. By ARTHUR M. CORWIN, A. M., M. D., Demonstrator of Physical Diagnosis in the Rush Medical College, Chicago; Attending Physician to the Central Free Dispensary, Department of Rhinology, Laryngology, and Diseases of the Chest. 200 pages. Illustrated. Cloth, flexible covers. Price, $1.25 net.

SYLLABUS OF OBSTETRICAL LECTURES in the Medical Department, University of Pennsylvania. By RICHARD C. NORRIS, A. M., M. D., Lecturer on Clinical and Operative Obstetrics, University of Pennsylvania. Third edition, thoroughly revised and enlarged. Crown 8vo. Price, Cloth, interleaved for notes, $2.00 net.

"This work is so far superior to others on the same subject that we take pleasure in calling attention briefly to its excellent features. It covers the subject thoroughly, and will prove invaluable both to the student and the practitioner. The author has introduced a number of valuable hints which would only occur to one who was himself an experienced teacher of obstetrics. The subject-matter is clear, forcible, and modern. We are especially pleased with the portion devoted to the practical duties of the accoucheur, care of the child, etc. The paragraphs on antiseptics are admirable; there is no doubtful tone in the directions given. No details are regarded as unimportant; no minor matters omitted. We venture to say that even the old practitioner will find useful hints in this direction which he cannot afford to despise."—*New York Medical Record.*

A SYLLABUS OF LECTURES ON THE PRACTICE OF SURGERY, arranged in conformity with "An American Text-Book of Surgery." By N. SENN, M. D., PH. D., Professor of Surgery in Rush Medical College, Chicago, and in the Chicago Polyclinic. Price, $2.00.

This, the latest work of its eminent author, himself one of the contributors to "An American Text-Book of Surgery," will prove of exceptional value to the advanced student who has adopted that work as his text-book. It is not only the syllabus of an unrivalled course of surgical practice, but it is also an epitome of or supplement to the larger work.

"The author has evidently spared no pains in making his Syllabus thoroughly comprehensive, and has added new matter and alluded to the most recent authors and operations. Full references are also given to all requisite details of surgical anatomy and pathology."—*British Medical Journal*, London.

AN OPERATION BLANK, with Lists of Instruments, etc. required in Various Operations. Prepared by W. W. KEEN, M. D., LL.D., Professor of Principles of Surgery in the Jefferson Medical College, Philadelphia. Price per Pad, containing Blanks for fifty operations, 50 cents net.

SECOND EDITION, REVISED FORM.

A convenient blank, suitable for all operations, giving complete instructions regarding necessary preparation of patient, etc., with a full list of dressings and medicines to be employed.

On the back of each blank is a list of instruments used—viz. general instruments, etc., required for all operations; and special instruments for surgery of the brain and spine, mouth and throat, abdomen, rectum, male and female genito-urinary organs, the bones, etc.

The whole forming a neat pad, arranged for hanging on the wall of a surgeon's office or in the hospital operating-room.

"Will serve a useful purpose for the surgeon in reminding him of the details of preparation for the patient and the room as well as for the instruments, dressings, and antiseptics needed."—*New York Medical Record*

"Covers about all that can be needed in any operation."—*American Lancet.*

"The plan is a capital one."—*Boston Medical and Surgical Journal.*

LABORATORY EXERCISES IN BOTANY. By EDSON S. BASTIN, M. A., Professor of Materia Medica and Botany in the Philadelphia College of Pharmacy. Octavo volume of 536 pages, 87 full-page plates. Price, Cloth, $2.50.

This work is intended for the beginner and the advanced student, and it fully covers the structure of flowering plants, roots, ordinary stems, rhizomes, tubers, bulbs, leaves, flowers, fruits, and seeds. Particular attention is given to the gross and microscopical structure of plants, and to those used in medicine. Illustrations have freely been used to elucidate the text, and a complete index to facilitate reference has been added.

"There is no work like it in the pharmaceutical or botanical literature of this country, and we predict for it a wide circulation."—*American Journal of Pharmacy.*

DIET IN SICKNESS AND IN HEALTH. By MRS. ERNEST HART, formerly Student of the Faculty of Medicine of Paris and of the London School of Medicine for Women; with an INTRODUCTION by Sir Henry Thompson, F. R. C. S., M. D., London. 220 pages; illustrated. Price, Cloth, $1.50.

Useful to those who have to nurse, feed, and prescribe for the sick. In each case the accepted causation of the disease and the reasons for the special diet prescribed are briefly described. Medical men will find the dietaries and recipes practically useful, and likely to save them trouble in directing the dietetic treatment of patients.

HOW TO EXAMINE FOR LIFE INSURANCE. By JOHN M. KEATING, M. D., Fellow of the College of Physicians and Surgeons of Philadelphia; Vice-President of the American Pædiatric Society; Ex-President of the Association of Life Insurance Medical Directors. Royal 8vo, 211 pages, with two large half-tone illustrations, and a plate prepared by Dr. McClellan from special dissections; also, numerous cuts to elucidate the text. Second edition. Price, Cloth, $2.00 net.

"This is by far the most useful book which has yet appeared on insurance examination, a subject of growing interest and importance. Not the least valuable portion of the volume is Part II., which consists of instructions issued to their examining physicians by twenty-four representative companies of this country. As the proofs of these instructions were corrected by the directors of the companies, they form the latest instructions obtainable. If for these alone, the book should be at the right hand of every physician interested in this special branch of medical science."—*The Medical News*, Philadelphia.

NURSING: ITS PRINCIPLES AND PRACTICE. By ISABEL ADAMS HAMPTON, Graduate of the New York Training School for Nurses attached to Bellevue Hospital; Superintendent of Nurses and Principal of the Training School for Nurses, Johns Hopkins Hospital, Baltimore, Md.; late Superintendent of Nurses, Illinois Training School for Nurses, Chicago, Ill. In one very handsome 12mo volume of 484 pages, profusely illustrated. Price, Cloth, $2.00 net.

This original work on the important subject of nursing is at once comprehensive and systematic. It is written in a clear, accurate, and readable style, suitable alike to the student and the lay reader. Such a work has long been a desideratum with those entrusted with the management of hospitals and the instruction of nurses in training-schools. It is also of especial value to the graduated nurse who desires to acquire a practical working knowledge of the care of the sick and the hygiene of the sick-room.

OBSTETRIC ACCIDENTS, EMERGENCIES, AND OPERATIONS. By L. CH. BOISLINIERE, M. D., late Emeritus Professor of Obstetrics in the St. Louis Medical College. 381 pages, handsomely illustrated. Price, $2.00 net.

"For the use of the practitioner who, when away from home, has not the opportunity of consulting a library or of calling a friend in consultation. He then, being thrown upon his own resources, will find this book of benefit in guiding and assisting him in emergencies."

INFANT'S WEIGHT CHART. Designed by J. P. CROZER GRIFFITH, M. D., Clinical Professor of Diseases of Children in the University of Pennsylvania. 25 charts in each pad. Price per pad, 50 cents net.

A convenient blank for keeping a record of the child's weight during the first two years of life. Printed on each chart is a curve representing the average weight of a healthy infant, so that any deviation from the normal can readily be detected.

THE CARE OF THE BABY. By J. P. CROZER GRIFFITH, M. D., Clinical Professor of Diseases of Children, University of Pennsylvania; Physician to the Children's Hospital, Philadelphia, etc. 392 pages, with 67 illustrations in the text, and 5 plates. 12mo. Price, $1.50.

A reliable guide not only for mothers, but also for medical students and practitioners whose opportunities for observing children have been limited.

"The whole book is characterized by rare good sense, and is evidently written by a master hand. It can be read with benefit not only by mothers, but by medical students and by any practitioners who have not had large opportunities for observing children."—*American Journal of Obstetrics.*

THE NURSE'S DICTIONARY of Medical Terms and Nursing Treatment, containing Definitions of the Principal Medical and Nursing Terms, Abbreviations, and Physiological Names, and Descriptions of the Instruments, Drugs, Diseases, Accidents, Treatments, Operations, Foods, Appliances, etc. encountered in the ward or in the sick-room. Compiled for the use of nurses. By HONNOR MORTEN, author of "How to Become a Nurse," "Sketches of Hospital Life," etc. 16mo, 140 pages. Price, Cloth, $1.00.

This little volume is intended for use merely as a small reference-book which can be consulted at the bedside or in the ward. It gives sufficient explanation to the nurse to enable her to comprehend a case until she has leisure to look up larger and fuller works on the subject.

DIET LISTS AND SICK-ROOM DIETARY. By JEROME B. THOMAS, M. D., Visiting Physician to the Home for Friendless Women and Children and to the Newsboys' Home; Assistant Visiting Physician to the Kings County Hospital; Assistant Bacteriologist, Brooklyn Health Department. Price, Cloth, $1.50 (Send for specimen List.)

One hundred and sixty detachable (perforated) diet lists for Albuminuria, Anæmia and Debility, Constipation, Diabetes, Diarrhœa, Dyspepsia, Fevers, Gout or Uric-Acid Diathesis, Obesity, and Tuberculosis. Also forty detachable sheets of Sick-Room Dietary, containing full instructions for preparation of easily-digested foods necessary for invalids. Each list is *numbered only*, the disease for which it is to be used in no case being mentioned, an index key being reserved for the physician's private use.

DIETS FOR INFANTS AND CHILDREN IN HEALTH AND IN DISEASE. By LOUIS STARR, M. D., Editor of "An American Text-Book of the Diseases of Children." 230 blanks (pocket-book size), perforated and neatly bound in flexible morocco. Price, $1.25 net.

The first series of blanks are prepared for the first seven months of infant life; each blank indicates the ingredients, but not the *quantities*, of the food, the latter directions being left for the physician. After the seventh month, modifications being less necessary, the diet lists are printed in full. *Formulæ* for the preparation of diluents and foods are appended.

SAUNDERS' NEW SERIES OF MANUALS

for Students and Practitioners.

THAT there exists a need for thoroughly reliable hand-books on the leading branches of Medicine and Surgery is a fact amply demonstrated by the favor with which the SAUNDERS NEW SERIES OF MANUALS have been received by medical students and practitioners and by the Medical Press. These manuals are not merely condensations from present literature, but are ably written by well-known authors and practitioners, most of them being teachers in representative American colleges. Each volume is concisely and authoritatively written and exhaustive in detail, without being encumbered with the introduction of "cases," which so largely expand the ordinary text-book. These manuals will therefore form an admirable collection of advanced lectures, useful alike to the medical student and the practitioner: to the latter, too busy to search through page after page of elaborate treatises for what he wants to know, they will prove of inestimable value; to the former they will afford safe guides to the essential points of study.

The SAUNDERS NEW SERIES OF MANUALS are conceded to be superior to any similar books now on the market. No other manuals afford so much information in such a concise and available form. A liberal expenditure has enabled the publisher to render the mechanical portion of the work worthy of the high literary standard attained by these books.

Any of these Manuals will be mailed on receipt of price (see next page for List).

W. B. SAUNDERS, Publisher,

925 Walnut Street, Philadelphia.

RECENT PUBLICATIONS

PENROSE'S DISEASES OF WOMEN

A Text-Book of Diseases of Women. By CHARLES B. PENROSE, M.D., PH.D., Professor of Gynecology in the University of Pennsylvania; Surgeon to the Gynecean Hospital, Philadelphia. Octavo volume of 529 pages, handsomely illustrated. Cloth, $3.50 net.

"I shall value very highly the copy of Penrose's "Diseases of Women" received. I have already recommended it to my class as THE BEST book."—HOWARD A. KELLY, *Professor of Gynecology and Obstetrics, Johns Hopkins University, Baltimore, Md.*

SENN'S GENITO-URINARY TUBERCULOSIS

Tuberculosis of the Genito-Urinary Organs, Male and Female. By NICHOLAS SENN, M.D., PH.D., LL.D., Professor of the Practice of Surgery and of Clinical Surgery, Rush Medical College, Chicago. Handsome octavo volume of 320 pages, illustrated. Cloth, $3.00 net.

SUTTON AND GILES' DISEASES OF WOMEN

Diseases of Women. By J. BLAND SUTTON, F.R.C.S., Assistant Surgeon to Middlesex Hospital, and Surgeon to Chelsea Hospital, London; and ARTHUR E. GILES, M.D., B.Sc. Lond., F.R.C.S. Edin., Assistant Surgeon to Chelsea Hospital, London. 436 pages, handsomely illustrated. Cloth, $2.50 net.

BUTLER'S MATERIA MEDICA, THERAPEUTICS, AND PHARMACOLOGY

A Text-Book of Materia Medica, Therapeutics, and Pharmacology. By GEORGE F. BUTLER, PH.G., M.D., Professor of Materia Medica and of Clinical Medicine in the College of Physicians and Surgeons, Chicago; Professor of Materia Medica and Therapeutics, Northwestern University, Woman's Medical School, etc. Octavo, 858 pages, illustrated. Cloth, $4.00 net; Sheep, $5.00 net.

SAUNDBY'S RENAL AND URINARY DISEASES

Lectures on Renal and Urinary Diseases. By ROBERT SAUNDBY, M.D. Edin., Fellow of the Royal College of Physicians, London, and of the Royal Medico-Chirurgical Society; Physician to the General Hospital; Consulting Physician to the Eye Hospital and to the Hospital for Diseases of Women; Professor of Medicine in Mason College, Birmingham, etc. Octavo volume of 434 pages, with numerous illustrations and 4 colored plates. Cloth, $2.50 net.

PYE'S BANDAGING

Elementary Bandaging and Surgical Dressing, with Directions Concerning the Immediate Treatment of Cases of Emergency. For the Use of Dressers and Nurses. By WALTER PYE, F.R.C.S., Late Surgeon to St. Mary's Hospital, London. Small 12mo, with over 80 illustrations. Cloth, flexible covers. Price, 75 cents net.

MALLORY AND WRIGHT'S PATHOLOGICAL TECHNIQUE

Pathological Technique. By FRANK B. MALLORY, A.M., M.D., Assistant Professor of Pathology, Harvard University Medical School; and JAMES H. WRIGHT, A.M., M.D., Instructor in Pathology, Harvard University Medical School. Octavo volume of 396 pages, handsomely illustrated. Cloth, $2.50 net.

"I have been looking forward to the publication of this book, and I am glad to say that I find it to be a most useful laboratory and post-mortem guide, full of practical information, and well up to date."—WILLIAM H. WELCH, *Professor of Pathology, Johns Hopkins University, Baltimore, Md.*

ANDERS' PRACTICE OF MEDICINE

A Text-Book of the Practice of Medicine. By JAMES M. ANDERS, M.D., PH.D., LL.D., Professor of the Practice of Medicine and of Clinical Medicine, Medico-Chirurgical College, Philadelphia. In one handsome octavo volume of 1287 pages, fully illustrated. Cloth, $5.50 net; Sheep or Half Morocco, $6.50 net.

JUST ISSUED

AN AMERICAN TEXT-BOOK OF GENITO-URINARY AND SKIN DISEASES

Edited by L. Bolton Bangs, M.D., Late Professor of Genito-Urinary and Venereal Diseases, New York Post-Graduate Medical School and Hospital; and William A. Hardaway, M.D., Professor of Diseases of the Skin, Missouri Medical College. Octavo volume of over 1200 pages, with 300 illustrations in the text, and 20 full-page colored plates. Prices: Cloth, $7.00 net; Sheep or Half Morocco, $8.00 net.

MOORE'S ORTHOPEDIC SURGERY

A Manual of Orthopedic Surgery. By James E. Moore, M.D., Professor of Orthopedics and Adjunct Professor of Clinical Surgery, University of Minnesota, College of Medicine and Surgery. 8vo, 356 pages, handsomely illustrated. Cloth, $2.50 net.

MACDONALD'S SURGICAL DIAGNOSIS AND TREATMENT

Surgical Diagnosis and Treatment. By J. W. Macdonald, M.D. Edin., L.R.C.S. Edin., Professor of the Practice of Surgery and of Clinical Surgery in Hamline University; Visiting Surgeon to St. Barnabas' Hospital, Minneapolis, etc. Octavo volume of 800 pages, handsomely illustrated. Cloth, $5.00 net; Half Morocco, $6.00 net.

CHAPIN ON INSANITY

A Compendium of Insanity. By John B. Chapin, M.D., LL.D., Physician-in-Chief, Pennsylvania Hospital for the Insane; late Physician-Superintendent of the Willard State Hospital, New York, etc. 12mo., 234 pages, illustrated. Cloth, $1.25 net.

KEEN ON THE SURGERY OF TYPHOID FEVER

The Surgical Complications and Sequels of Typhoid Fever. By Wm. W. Keen, M.D., LL.D., Professor of the Principles of Surgery and of Clinical Surgery, Jefferson Medical College, Philada. Octavo volume of 400 pages. Cloth, $3.00 net.

VAN VALZAH AND NISBET'S DISEASES OF THE STOMACH

Diseases of the Stomach. By William W. van Valzah, M.D., Professor of General Medicine and Diseases of the Digestive System and the Blood, New York Polyclinic; and J. Douglas Nisbet, M.D., Adjunct Professor of General Medicine and Diseases of the Digestive System and the Blood, New York Polyclinic. Octavo volume of 674 pages, illustrated. Cloth, $3.50 net.

IN PREPARATION

AN AMERICAN TEXT-BOOK OF DISEASES OF THE EYE, EAR, NOSE, AND THROAT

Edited by G. E. de Schweinitz, M.D., Professor of Ophthalmology in the Jefferson Medical College; and B. Alexander Randall, M.D., Professor of Diseases of the Ear in the University of Pennsylvania and in the Philadelphia Polyclinic.

CHURCH AND PETERSON'S NERVOUS AND MENTAL DISEASES

Nervous and Mental Diseases. By Archibald Church, M.D., Professor of Mental Diseases and Medical Jurisprudence, Northwestern University Medical School, Chicago; and Frederick Peterson, M.D., Clinical Professor of Mental Diseases, Woman's Medical College, New York, etc.

KYLE ON THE NOSE AND THROAT

Diseases of the Nose and Throat. By D. Braden Kyle, M.D., Clinical Professor of Laryngology and Rhinology, Jefferson Medical College, Philadelphia; Consulting Laryngologist, Rhinologist, and Otologist, St. Agnes' Hospital, etc.

STENGEL'S PATHOLOGY

A Manual of Pathology. By Alfred Stengel, M.D., Physician to the Philadelphia Hospital; Professor of Clinical Medicine in the Woman's Medical College; Physician to the Children's Hospital, etc.

HIRST'S OBSTETRICS

A Text-Book of Obstetrics. By Barton Cooke Hirst, M.D., Professor of Obstetrics, University of Pennsylvania.

HEISLER'S EMBRYOLOGY

A Text-Book of Embryology. By John C. Heisler, M.D., Professor of Anatomy, Medico-Chirurgical College, Philadelphia.